Transforming Agriculture with Emerging Technologies for a Sustainable Future

This book presents a discussion of advanced novel technologies and their integration with intelligent sensor data processing. It focuses on optimization models for agriculture issues such as food traceability, supply chain management, and energy resource management. The text also covers case studies from agriculture-based companies to provide more practical exposure to the readers.

Key Features:

- Discusses smart architectures through a blend of technologies such as the Internet of Things, edge computing, fog computing, and blockchain for agriculture data handling and automation.
- Covers approaches toward supply chain management, food traceability, health ecosystems, and energy resource management.
- Presents solutions and approaches for real-time issues such as livestock monitoring, agriculture waste management, crop health, and irrigation issues.
- Highlights smart optimized solutions through novel farming technologies such as vertical farming, Unmanned Aerial Vehicles (UAV)-based farming solution, and agriculture cybernetics.
- Showcases smart strategies for agricultural waste management and explains crop health sensing methods and applications.

It is primarily written for senior undergraduates, graduate students, and academic researchers in the fields of electrical engineering, electronics and communication engineering, computer science and engineering, agricultural science, and information technology.

Smart Engineering Systems: Design and Applications
Series Editor – Suman Lata Tripathi

Internet of Things: Robotic and Drone Technology
Edited by Nitin Goyal, Sharad Sharma, Arun Kumar Rana, and Suman Lata Tripathi

Smart Electrical Grid System: Design Principle, Modernization, and Techniques
Edited by Krishan Arora, Suman Lata Tripathi, and Sanjeevikumar Padmanaban

Artificial Intelligence, Internet of Things (IoT) and Smart Materials for Energy Applications
Edited by Mohan Lal Kolhe, Kailash J. Karande, and Sampat G. Deshmukh

Artificial Intelligence for Internet of Things: Design Principle, Modernization, and Techniques
Edited by N. Thillaiarasu, Suman Lata Tripathi, and V.Dhinakaran

Machine Learning and Internet of Things in Solar Power Generation
Edited by Prabha Umapathy, Jude Hemanth, Shelej Khera, Abinaya Inbamani, and Suman Lata Tripathi

Microgrid: Design, Optimization, and Applications
Edited by Amit Kumar Pandey, Sanjeevikumar Padmanaban, Suman Lata Tripathi, Vivek Patel, and Vikas Patel

Transforming Agriculture with Emerging Technologies for a Sustainable Future
Shanu Sharma, Ayushi Prakash, Sandhya Avasthi, and Vijayan Sugumaran

For more information about this series, please visit: https://www.routledge.com/Smart-Engineering-Systems/book-series/CRCSESDA

Transforming Agriculture with Emerging Technologies for a Sustainable Future

Edited by
Shanu Sharma, Ayushi Prakash,
Sandhya Avasthi, and Vijayan Sugumaran

CRC Press
Taylor & Francis Group
Boca Raton London New York

CRC Press is an imprint of the
Taylor & Francis Group, an **informa** business

Front cover image: Shutterstock

First edition published 2026
by CRC Press
2385 NW Executive Center Drive, Suite 320, Boca Raton FL 33431

and by CRC Press
4 Park Square, Milton Park, Abingdon, Oxon, OX14 4RN

CRC Press is an imprint of Taylor & Francis Group, LLC

ISBN: 978-1-032-79967-4 (hbk)
ISBN: 978-1-032-79968-1 (pbk)
ISBN: 978-1-003-49478-2 (ebk)

DOI: 10.1201/9781003494782

Typeset in Sabon LT Std
by codeMantra

Contents

Preface

TRANSFORMING AGRICULTURE WITH EMERGING TECHNOLOGIES FOR A SUSTAINABLE FUTURE

In the past few decades, the Smart Agriculture domain has attracted extraordinary attention from academia, government, industries, and various agricultural communities. Furthermore, a huge transformation has been observed with the emergence of novel technologies such as advanced machine learning, that is, generative AI, Large Language Models (LLMs), real-time integration of IoT, blockchain-enabled secured infrastructure, and so forth. These emerging technologies have the capability of addressing various challenges in the agriculture industry, including a lack of assured irrigation, inadequate demand, and prediction, and overuse or misuse of fertilizers and pesticides. They can create a significant global impact on agricultural productivity at all levels of the value chain.

Due to the remarkable power of novel technologies in changing and modernizing the agricultural value chain, this book is intended to provide recent advancements and innovations toward novel strategies, mechanisms, and devices for future agriculture automation and transforming industrial functions to improve agricultural industrial efficiency and safety for a sustainable future.

This book is an edited collection of cutting-edge chapters in the area of integrating novel technologies for developing approaches for sustainable agriculture. Featuring insights from academics, technologists, and field experts, topics range from AI and machine learning, IoT and blockchain, and 5G and intelligent automation to biosurfactant optimization. Every chapter explores the theoretical design and development of these technologies, and the narratives run through real-life use cases, as well as insights into future-forward innovations.

This book begins with a foundational understanding of the key pillars of sustainable agriculture for the future in Chapter 1, establishing the baseline to explore opportunities for innovation.

Chapters 2–6 focus on a discussion of various emerging technologies like AI, ML, IoT, Blockchain, and 5G. These chapters explore how these technologies are capable of transforming the agriculture sector through productivity enhancement and resource management. For instance, the integration of IoT and deep learning models can automate the disease detection of plants, while 5G-enabled solutions have the power to resolve the logistical issues in precision agriculture.

Chapters 7–9 highlight the transformative role of intelligent automation, IoT, and AI-driven innovations in advancing sustainable agriculture. From fuzzy wireless sensor networks enhancing precision in Indian farming to IoT-based smart farming systems and AI-optimized biosurfactant production from agricultural waste, they showcase technology's potential to boost efficiency, reduce environmental impact, and support resource sustainability.

AI, ML, and IoT-powered solutions in addressing critical agricultural challenges are presented in Chapters 10–13. From early plant disease detection and shelf-life prediction to crop yield estimation and wild animal tracking, they showcase how predictive and intelligent technologies enhance productivity, minimize losses, and support data-driven decision-making for a smarter, more secure agricultural future.

This book is prepared to serve as a reference publication for researchers, policymakers, agronomists, tech developers, and students. It serves not only as a comprehensive review but also as a guide for the present scope of technology integration in agriculture and for the future prospects to explore. The long-term ambition is to promote integrated, multi-disciplinary approaches that will lead to the next generation of smart, sustainable agriculture for everyone.

We thank all the contributors and reviewers for their invaluable time, expertise, and insights. It is their passion that makes possible the publication that seeks to both educate and inspire, and to catalyze the inventions needed to feed the future, sustainably.

Dr. Shanu Sharma
Dr. Ayushi Prakash
Dr. Sandhya Avasthi
Dr. Vijayan Sugumaran
Editors

About the editors

Shanu Sharma is an Associate Professor in the Department of Computer Science and Engineering at ABES Engineering College, Ghaziabad (Affiliated to A.K.T. University, Lucknow). She holds a Ph.D. from Amity University, Uttar Pradesh, and an M.Tech. in Intelligent Systems from IIIT Allahabad. With over 14 years of teaching and research experience, she has taught a variety of undergraduate and postgraduate courses, including Image Processing, Data Mining, Machine Learning, Data Science, Data Structures, Analysis and Design of Algorithms, and Compiler Design. Her research interests span cognitive computing, computer vision, pattern recognition, and machine learning. She has contributed to numerous reputed journals and conferences and is actively involved with several international journals and conferences as a reviewer and guest editor. She has been actively engaged in editing technical books in the fields of artificial intelligence, intelligent systems, and emerging technologies. These books contribute to advancing industry-relevant knowledge and are widely used by researchers, practitioners, and students. She has also curated special issues in international journals focused on interdisciplinary and cutting-edge topics. She has also served as a conference organizer and session chair for several international conferences, contributing to the academic and research community by facilitating knowledge exchange and collaborative opportunities. She is a senior member of IEEE and an active member of multiple professional societies, including ACM, Soft Computing Research Society (SCRS), EUSFLAT, and IAENG.

Ayushi Prakash is a Professor in the Department of Computer Science and Engineering at Ajay Kumar Garg Engineering College, Ghaziabad, Dr. A.P.J. Abdul Kalam Technical University, Uttar Pradesh, where she has been since September 2023. She received a B.Tech. from Dr. R.M.L. Awadh University in 2009 and an M.Tech. from the AKTU State University of Uttar Pradesh at Lucknow. She received her Ph.D. in Computer Science and Engineering from Dr. K. N. Modi University in 2022. From 2010 to 2024, she also has a continuous contribution in the Training and Placement Department at Dr. A.P.J. Abdul Kalam Technical University, Uttar Pradesh. Her research interests span both information retrieval and AIML. She is a senior member of IEEE and a professional lifetime member of the Indgiants Association of Educators and Professional Trainers. She has received many best paper awards in international conferences. She has contributed to many international conferences as an organizer, advisor, session chair, and reviewer. She has more than 20 Sci/Scopus-indexed research papers and 20 research papers in renowned national and international journals and conferences. She has published many patents, chapters, and research articles.

Sandhya Avasthi is an Associate Professor in the Department of Computer Science and Engineering at ABES Engineering College (affiliated with Dr. A.P.J. Abdul Kalam Technical University), Ghaziabad, India. She earned her Ph.D. in Machine Learning and Text Mining from Amity University, Noida. She holds an M.Tech. in Computer Science and Engineering from Uttar Pradesh Technical University, Lucknow, and a B.E. in Computer Science and Engineering from Dr. B.R. Ambedkar University, Agra. With over 19 years of teaching experience, Dr. Avasthi is an active researcher in the domains of machine learning and data mining. Her research interests include natural language processing, information extraction, information retrieval, data science, and business intelligence. She has published numerous research papers in peer-reviewed international journals, conference proceedings, and edited book volumes. In addition to her academic responsibilities, Dr. Avasthi contributes actively to the research community. She serves as a reviewer for reputed journals and conferences. She is also a member of professional bodies such as IEEE and ACM and remains consistently engaged in academic and professional development activities.

Vijayan Sugumaran is a Professor of Management Information Systems and Chair of the Department of Decision and Information Sciences at Oakland University, Rochester, Michigan, USA. He received his Ph.D in Information Technology from George Mason University, Fairfax, USA. His research interests are in the areas of big data analytics, ontologies and semantic web, intelligent agent and multi-agent systems, and component-based development. He has published over 200 peer-reviewed articles in journals, conferences, and books. He has edited 12 books and serves on the Editorial Board of 8 journals. He has published in top-tier journals such as *Information Systems Research*, *ACM Trans on Database Systems*, *IEEE Trans on Engineering Management*, *Communications of the ACM*, and *IEEE Software*. He is the editor-in-chief of the *International Journal of Intelligent Information Technologies*. He is the Chair of the Intelligent Agent and Multi-Agent Systems mini-track for the Americas Conference on Information Systems (AMCIS 1999–2019). He has served as the program co-chair for the *International Conference on Applications of Natural Language to Information Systems* (NLDB 2008, 2013, 2016, and 2019).

Contributors

Mohamed A. E. Abdel Rahman
Division of Environmental Studies and Land Use
National Authority for Remote Sensing and Space Sciences (NARSS)
Cairo, Egypt

Abhijith R. P.
Marian Engineering College
Kerala, India

Salman Khursheed Ahmad
Galgotias University
Greater Noida, India

Jayashree S. Awati
Rajarambapu Institute of Technology
Rajaramnagar, Sakharale, Affiliated to Shivaji University
Kolhapur, India

Alok Bajoria
MIT World Peace University
Pune, India

Darshit Bansal
ABES Engineering College
Ghaziabad, India

Sandeep Bhatia
Galgotias University
Greater Noida, India

Mahi Bohara
MIT World Peace University
Pune, India

Saransh Chauhan
ABES Engineering College
Ghaziabad, India

Ankit Dahiya
MIT World Peace University
Pune, India

Showkat A. Dar
GITAM School of Technology
GITAM Deemed to be University
Bengaluru, India

Nagaraj V. Dharwadkar
Central University of Karnataka
Kalaburgi, India

Siddhi Dingorkar
MIT World Peace University
Pune, India

H. S. Ganesha
JSS University
Noida, India

Km Ikra
Galgotias University
Greater Noida, India

R. Indhumathi
Idhaya College for Women
Bharathidasan University
Kumbakonam, India

Indu
USICT, Gautam Buddha University
Greater Noida, India

Malik Shahzad Ahmed Iqbal
Acharya University Karakul
Stantslya-Karakul, Uzbekistan

Zainul Abdin Jaffery
Jamia Millia Islamia
Delhi, India

Vishal Jayaswal
Ajay Kumar Garg Engineering College
Ghaziabad, India

Sharma Ji
Ajay Kumar Garg Engineering College
Ghaziabad, India

Vijay H. Kalmani
Rajarambapu Institute of Technology
Rajaramnagar, Sakharale, Affiliated to Shivaji University
Kolhapur, India

Sonali Kokale
Department of Agronomy, ITM University
Gwalior, India

Pranali Kosamkar
MIT World Peace University
Pune, India

Atin Kumar
School of Agriculture
Uttaranchal University
Dehradun, India

Mahesh Kumbar
Rajarambapu Institute of Technology
Rajaramnagar, Sakharale, Affiliated to Shivaji University
Kolhapur, India

Shabana Mehfuz
Jamia Millia Islamia
Delhi, India

Chandrahas Mishra
ABES Engineering College
Ghaziabad, India

Tribhuvan Mishra
ABES Engineering College
Ghaziabad, India

Vaishnavi Mishra
ABES Engineering College
Ghaziabad, India

C. Mithra
School of Computing (SOC)
Vel Tech Rangarajan Dr. Sagunthala R&D Institute of Science and Technology
Chennai, India

Nishant Kumar Pathak
Ajay Kumar Garg Engineering College
Ghaziabad, India

Raj Pooniya
Department of Computer Science and Engineering
Meerut Institute of Engineering and Technology
Meerut, India

Ayushi Prakash
Ajay Kumar Garg Engineering College
Ghaziabad, India

Rani M. R.
Government Engineering College Palakkad
Kerala, India

Aafaq A. Rather
Symbiosis Statistical Institute
Symbiosis International Deemed University
Pune, India

Sharad Sachan
School of Agriculture
Lovely Professional University
Phagwara, India

T. Saravanan
GITAM School of Technology
GITAM Deemed to be University
Bengaluru, India

Yukta Shah
MIT World Peace University
Pune, India

Shanu Sharma
ABES Engineering College
Ghaziabad, India

Vandita Sharma
Meerut Institute of Engineering and Technology, Meerut
Affiliated to CCS University
Meerut, India

Abhishek Kumar Shukla
ABES Engineering College
Ghaziabad, India

Pushkal Kumar Shukla
Ajay Kumar Garg Engineering College
Ghaziabad, India

Simi M. R.
Mahaguru Institute of Technology
Kerala, India

Aditya Pratap Singh
Ajay Kumar Garg Engineering College
Ghaziabad, India

Smriti Singh
GL Bajaj Institute of Technology and Management
Greater Noida, India

Vinay Singh
ABES Engineering College
Ghaziabad, India

Abhay Singhal
ABES Engineering College
Ghaziabad, India

Akhilesh Kumar Srivastava
ABES Engineering College
Ghaziabad, India

Asmit Yadav
ABES Engineering College
Ghaziabad, India

Chapter 1

Key pillars of sustainable agriculture for the future

Simi M.R., Abhijith R.P., and Rani M.R.

1.1 INTRODUCTION

The growing global population presents a notable challenge to food security, necessitating substantial enhancements in agricultural productivity. However, heavy reliance on chemical fertilizers presents environmental and health risks. As the global population is projected to reach 9.1 billion by 2050, there is a pressing need to explore sustainable farming practices capable of meeting the food demands of an additional two billion people. Agroecological cropping practices in temperate regions offer a promising avenue for achieving this goal while minimizing environmental impact [1]. Agriculture is intimately connected to the environment. Given its notable consequences on the environment, attaining sustainable agricultural development is crucial for present and future generations. Agriculture is experiencing unprecedented challenges in the human age, requiring a transition toward sustainability to prevent self-ruin [2]. Traditional agriculture is increasingly insufficient in tackling the complicated issues of food production and ecosystem preservation, demanding a transition toward sustainable agricultural practices. In agriculture, environmental sustainability includes responsibly handling natural systems and resources that farms rely on [3].

Sustainable agriculture includes methods that meet the needs of society both now and in the future [4]. In other words, sustainability in agriculture is grounded in the principle of fulfilling current needs without compromising the ability of future generations to accomplish their own, amidst a multitude of global challenges spanning from hunger and obesity to climate change and soil degradation. Sustainable agriculture is often characterized by its three main goals: environmental health, economic profitability, and social and economic equity, reflecting a holistic approach to farming practices [5]. Sustainable agriculture promotes economic growth and market competitiveness while meeting basic societal needs without harming the environment or depleting non renewable resources [6,7].

Agriculture can both harm and benefit the environment. The primary challenge for the agricultural sector is to provide food for an increasing global population while reducing environmental impacts and preserving natural

DOI: 10.1201/9781003494782-1

resources for future generations. Cyclic interactions between water (irrigation), food (agriculture), and energy have caused severe socio-environmental crises in India. These include groundwater depletion, energy shortages, irrigation system failures, food insecurity, and livelihood losses [8]. Addressing these issues requires targeted policy interventions [9,10]. There is an inconsistency in approaches to agricultural sustainability, as various policies address different aspects of sustainability without a unified strategy [11,12]. The concept of sustainable agriculture helps identify the primary environmental impacts of agriculture and the major environmental challenges it faces. Sustainable agriculture offers a holistic approach to food production, integrating various scientific disciplines and prioritizing ecological integrity and social responsibility over mere productivity and profit [2]. Sustainable agriculture presents a viable alternative for addressing pressing agronomic challenges while ensuring environmental sustainability and human well-being, emphasizing the need for interdisciplinary collaboration and innovation. The Sustainability Assessment of Farming and the Environment (SAFE) framework was introduced by [13], offering a structured approach to assessing sustainability in agricultural systems. The framework encompasses principles, criteria, indicators, and reference values, providing a hierarchical structure for evaluation at parcel, farm, and broader spatial levels. It highlights the versatile nature of ago-ecosystems, incorporating environmental, economic, and social considerations. While not recommending a global solution, the SAFE framework acts as a practical tool for recognizing, enhancing, and assessing agricultural practices and policies in attunement with sustainability goals.

The article is organized to provide a thorough examination of the key pillars for sustainable agriculture in the future. It starts with an introduction to the foundational pillars of sustainable agriculture, followed by a comprehensive discussion on environmental health, underscoring the importance of protecting natural resources and ecosystems. Next, the attention changes to economic viability, dealing with the need for long-term profit potential for farmers and stakeholders, and social and economic equity, which showcases fair availability to resources and opportunities for all. The article additionally analyzes resource efficiency, assuring efficient utilization of water, energy, and land, and biodiversity conservation, which stresses the necessity to protect and sustain ecosystem services. The part of innovation and knowledge transfer is reviewed as a catalyst for embracing new sustainable practices, establishing a path for the investigation of future directions in sustainable agriculture, containing budding trends and challenges. The following sections look into technological advancements, such as precision farming and biotechnology, and climate-resilient practices, which empower adjustment to climate change consequences. The article also stresses the importance of sustainable soil management, maintaining soil health for long-term productivity, and introduces the concept of a circular economy in agriculture, which reduces waste and encourages resource reuse. Lastly, the part of policy and institutional aid is underscored as a driver for carrying out sustainable practices, and the article wraps up with a blending of these key elements, providing avenues for a sustainable agricultural future.

1.2 IDENTIFYING THE KEY PILLARS OF SUSTAINABLE AGRICULTURE

In light of numerous issues like climate change, biodiversity loss, soil degradation, water resource depletion, and economic inequality, it is crucial to ensure agricultural sustainability. Furthermore, the very methods that have kept agriculture afloat in recent decades have greatly exacerbated these issues, underscoring the pressing need for revolutionary strategies to protect ecological integrity and food security [14,15].

A number of fundamental principles underpin sustainable agriculture, guaranteeing its long-term sustainability and efficacy. Utilizing resource-saving techniques and technology that can support less intensive farming methods and strengthen natural capital is what is meant by enhanced sustainability. A multidimensional strategy to farming, sustainable agriculture seeks to satisfy present-day societal expectations while guaranteeing that future generations can likewise satisfy theirs. Three main pillars—economic viability, environmental health, and social equity—are at the heart of this strategy. Early criticisms of the economic status quo and attempts to balance economic expansion with social and ecological concerns led to the gradual development of the now widely accepted three-pillar concept of sustainability, which includes social, economic, and environmental aspects [16]. Although the three-pillar model is now widely used in discussions of sustainability, there are still issues with its theoretical foundations.

By guaranteeing farming's continued profitability, economic viability gives farmers a steady income and supports the overall economy. This entails implementing procedures and tools that increase output and effectiveness without using up too many resources. The goal of environmental health is to preserve and improve the natural ecosystem. This entails lowering pollutants and greenhouse gas emissions, conserving water, soil, and biodiversity, and enhancing the quality of the air, water, and land.

Social equity focuses on the fair distribution of the benefits and responsibilities of agriculture. It ensures that all members of society, including marginalized and vulnerable populations, have access to resources, opportunities, and a voice in decision-making processes related to agriculture. Together, these pillars create a balanced approach to farming that supports the well-being of people, the planet, and the economy, laying the foundation for a sustainable future.

1.3 ENVIRONMENTAL HEALTH

Long-term prosperity and the shift to a sustainable economic model, especially in the agriculture sector, depend heavily on environmental responsibility. A key component of sustainable agriculture is environmental health, which emphasizes the protection and improvement of natural resources that are essential to ecological balance and agricultural output. Even with initiatives to increase agricultural output, the majority of attention is still on intensification techniques rather than on adopting the comprehensive strategy of agroecology, which opposes the traditional agro-industrial system. In contrast to simple modifications within the current neoliberal agricultural model, agroecology provides a revolutionary alternative to conventional agriculture by promoting sustainable practices that prioritize biodiversity, lessen reliance on outside inputs, and foster self-sufficiency [17].

Preserving biodiversity, saving water, and preserving soil fertility are important elements [3,10]. Crop development depends on healthy soil, and sustainable techniques like crop rotation, cover crops, and decreased tillage help to improve the soil structure, stop soil erosion, and promote nutrient cycling. Efficient irrigation technologies, such as drip irrigation and rainwater harvesting, reduce water waste and preserve water quality, making water conservation equally important. The limitations and disadvantages of chemical fertilizers in agriculture may be addressed with biofertilizers, which capitalize on the potential of beneficial microorganisms. These microorganisms, which include fungi and bacteria that promote plant growth, have shown promise in increasing crop yields and fostering environmental sustainability [18].

Because it supports ecosystem processes like pollination, insect control, and nutrient cycling, biodiversity is essential to successful agriculture. Agroforestry, preserving natural ecosystems, and incorporating a variety of crop species are among practices that promote a rich biodiversity that increases resistance to pests, illnesses, and climate change. All things considered, environmental health in sustainable agriculture supports both ecological integrity and agricultural productivity for future generations by safeguarding the natural resources that farming systems rely on. Cultivating environmental responsibility, a fundamental component of sustainable development and corporate social responsibility, requires raising environmental awareness. In order to ensure a more seamless transition to sustainable practices, research on agri-environmental measures in Slovakia from 2007 to 2013 emphasizes the need for new initiatives to increase environmental awareness and responsibility in agriculture [6]. By improving soil properties and characteristics, adopting sustainable soil management techniques increases crop yields, increases the availability of water and nutrients, and decreases soil erosion, all of which support environmental preservation and food security [19]. Enhancing soil health and reducing environmental degradation brought on by urbanization and population expansion need sustainable soil management, especially through conservation tillage methods and Conservation Agriculture systems.

Agovino et al. [4] examined the intricate connection between climate change and agriculture, emphasizing the ways in which sustainable practices can both guarantee stable agricultural output and lessen environmental effects. A promising strategy for tackling the problems caused by climate change while improving agricultural output and the standard of living for marginalized groups is climate-smart agriculture (CSA) [20]. Despite its promise to advance sustainable agriculture and enhance global food security, there are obstacles to successfully putting CSA practices into practice, such as a lack of knowledge and inadequate field-level integration of its components. Realizing the advantages of CSA practices and ensuring a brighter future for agriculture depend on efforts to spread them via research and educational programs.

One noteworthy example of how environmental health is included in sustainable agriculture is Andhra Pradesh's "Zero Budget Natural Farming" (ZBNF) program. An agroecological technique called ZBNF seeks to lessen farmers' reliance on imported inputs and encourage organic farming practices that improve soil health, save water, and preserve biodiversity. With the help of the state government, ZBNF—which was started by Subhash Palekar—grew significantly in popularity in Andhra Pradesh with the goal of enhancing farmers' livelihoods while maintaining environmental sustainability. Using locally available natural inputs, mulching the soil with crop leftovers or organic matter, intercropping, and agroforestry are some of ZBNF's main activities. Notwithstanding the difficulties and obstacles, the need for healthier and more ecologically friendly food options among consumers, as well as increased knowledge of the advantages of organic farming in addressing sustainability issues, have propelled organic farming's notable global expansion [21]. Social paradigm and value changes are necessary to preserve soil health and natural resources for future generations, with a focus on the significance of giving sustainable practices and environmental stewardship in agricultural systems first priority [22].

1.4 ECONOMIC VIABILITY

Significant obstacles to agricultural sustainability include soil loss, rising water demand, pollution, biodiversity loss, and greenhouse gas emissions. To reduce environmental harm and protect natural resources, alternative methods must be adopted. The urgent need to feed the world's expanding population while using the least amount of land, water, and energy possible emphasizes how crucial it is to develop technical solutions and deal with socioeconomic problems in order to guarantee agriculture's long-term viability. Technical advancements are important, but in order to achieve sustainable agricultural practices and solve future food security issues, socioeconomic problems like population growth, resource disparity, and access to education must also be addressed.

A key component of sustainable agriculture is economic viability, which guarantees that farming methods are long-term productive and financially viable. For farmers to ensure steady revenues, this entails optimizing for resource efficiency, cutting expenses, and enhancing market accessibility. Farmers may maximize the use of inputs like water, fertilizer, and pesticides, cutting waste and production costs, by implementing sustainable methods like precision agriculture. Additionally, diversifying livestock and crops helps reduce the risk of unfavorable weather and market swings, resulting in more reliable sources of income.

Fair market accessibility is essential for economic sustainability. This can be accomplished through cooperative marketing, fair trade programs, and local food systems, all of which assist farmers in getting fair pricing for their goods. Financial instruments like financing and insurance can also help farmers invest in sustainable practices and manage risks. In addition to sustaining the livelihoods of farmers and rural communities, ensuring the economic viability of agriculture promotes the ongoing adoption of sustainable practices, which enhances the agricultural sector's resilience and sustainability overall.

1.5 SOCIAL AND ECONOMIC EQUITY

Evaluating social sustainability in agriculture is crucial, particularly in light of the social importance of farming and the current emphasis on environmental improvement by policymakers. Globally, farms are embracing the idea of sustainability, underscoring the need for a common comprehension of the social aspect of sustainability in agriculture [23]. The social pillar of sustainability in agriculture frequently includes elements that are difficult to quantify since farms differ in many ways and there is a lack of common information. A key component of sustainable agriculture is social and economic fairness, which emphasizes treating all parties in the agricultural industry fairly and providing them with equal opportunity. By guaranteeing equitable access to resources and opportunities, safe working conditions, and fair salaries, this pillar focuses on enhancing the well-being of farmers, laborers, and rural communities. Sustainable agriculture aims to help marginalized and smallholder farmers by giving them access to markets, loans, land, and education. This contributes to poverty reduction and the development of resilient communities. In order to promote economic justice, fair trade policies and cooperative models can guarantee that farmers and workers receive an equitable portion of the profits.

Furthermore, because women play a vital part in farming and food production, sustainable agriculture promotes gender equality and women's empowerment. Sustainable agriculture contributes to the development of a more just and equitable food system that benefits all parties—from producers to consumers—by tackling social injustices and promoting inclusive growth. It is imperative to develop strategies that get around the difficulties associated with qualitative evaluation, emphasizing the necessity of regular technique updates to improve information quality and correspond with changing rural social issues.

Assessing the social effects of farms' operations on local communities directly may provide insightful information about social sustainability and point to areas where assessment techniques need to be improved. A thorough validation of the social aspect of sustainability might be obtained by combining social assessments with assessments of environmental and economic performance. This could improve comprehension through comparative study across various geographic contexts [24].

The Self-Employed Women's Association (SEWA) program in Gujarat, India, is a powerful case study that emphasizes social and economic fairness in sustainable agriculture. A labor union called SEWA advocates for women who work in the unorganized sector, such as in agriculture. Enhancing underprivileged women farmers' access to markets, resources, and sustainable farming methods is the main goal of SEWA. The goal of this program is to improve their social and economic position in their communities. In addition to improving the livelihoods of women farmers, the initiative fosters more resilient and inclusive agricultural communities by empowering them.

1.6 RESOURCE EFFICIENCY

A key component of sustainable agriculture is resource efficiency, which emphasizes the best possible use of inputs like energy, water, and nutrients to increase output while reducing waste and environmental damage. The economic and environmental sustainability of agricultural activities is guaranteed by effective resource management. By getting water straight to the roots of plants and collecting rainfall for later use, methods like drip irrigation and rainwater harvesting help save water. With the use of precision agriculture technologies, such as soil sensors and GPS-guided equipment, farmers can apply pesticides and fertilizers more precisely, lowering the amount required and avoiding runoff into waterways.

The critical need for resource-efficient farming systems is highlighted by the growing global concern over the environmental impact of contemporary agriculture, which is made worse by population increase and rising demands for agricultural products [14,19,25]. In particular, there are serious environmental problems associated with intensively managed agricultural systems' inefficient use of nitrogen, such as soil erosion and water contamination. To overcome these obstacles, we must move toward extremely resource-efficient systems that strike a balance between societal acceptance, ecological safety, and profitability. Opportunities to improve crops and farming systems' nitrogen-use efficiency (NUE) must be carefully considered and prioritized in order to meet this goal. This calls for a multidisciplinary strategy that includes sustainable farming methods, optimal input management, and genetic improvement. In order to guarantee sustainable resource management at the local and global levels, the creation of ecologically friendly agricultural systems ultimately necessitates a coordinated effort incorporating quantitative systems research, the use of best practices, and supportive legislation.

In order to lessen reliance on fossil fuels, sustainable agriculture encourages the use of renewable energy sources like solar and wind power. Energy efficiency is also crucial. Furthermore, techniques like crop rotation and conservation tillage enhance soil health and nutrient cycling, reducing the demand for synthetic fertilizers and increasing soil productivity. In addition to lowering expenses and raising yields, sustainable agriculture also lessens farming's environmental impact by emphasizing resource efficiency, which helps ensure the long-term viability of agricultural systems.

1.7 BIODIVERSITY CONSERVATION

A key component of sustainable agriculture is biodiversity conservation, which acknowledges the value of preserving a variety of habitats to promote ecological balance, resilience, and agricultural output. By encouraging habitat preservation, genetic diversity in crops and livestock, and ecological restoration initiatives, sustainable agricultural techniques seek to protect biodiversity. Pollinators, beneficial insects, and wildlife depend on the preservation of natural ecosystems, including woods, wetlands, and grasslands in agricultural settings. These habitats sustain ecosystem services, including soil fertility, pest control, and water regulation, that are critical to agriculture's sustainability.

In order to increase resilience against pests, illnesses, and shifting environmental conditions, sustainable agriculture also highlights the significance of maintaining genetic variety in crops and cattle. While maintaining traditional knowledge and cultural history, crop diversity through techniques like intercropping and heirloom cultivars helps reduce the hazards associated with monoculture farming. In addition to promoting ecosystem health, sustainable agriculture increases agricultural production, resilience, and food security for present and future generations by placing a high priority on biodiversity protection.

1.8 INNOVATION AND KNOWLEDGE TRANSFER

With an emphasis on the ongoing creation and sharing of novel techniques, technologies, and insights that improve agricultural sustainability, innovation, and knowledge transfer are essential pillars of sustainable agriculture.

To increase output, efficiency, and environmental stewardship, these pillars guarantee that farmers and other agricultural stakeholders have access to the newest developments and best practices.

The creation of crop types resistant to drought, precision farming technologies, and renewable energy sources are examples of innovation in sustainable agriculture. Farmers are better able to use resources efficiently, lessen their influence on the environment, and adjust to the changing climate thanks to these advancements. For instance, precision agriculture improves yields and consumes less water and fertilizer by customizing agricultural methods to particular field circumstances using data and technology.

Farmers, researchers, policymakers, and extension agencies must effectively communicate and share these innovations and best practices in order for knowledge transfer to take place. Extension services, research partnerships, farmer training programs, and the use of social media and digital platforms can all help achieve this. Knowledge transfer assists farmers in implementing sustainable practices that are suited to their particular situations by promoting a culture of ongoing learning and adaptation.

In order to ensure that the advantages of scientific and technological advancements reach all levels of the agricultural community and ultimately contribute to the resilience and sustainability of global food systems, innovation and knowledge transfer work together to drive the adoption of sustainable agricultural practices. Through the controlled distribution of nutrients to crops, nano-biotechnology is emerging as a ground-breaking technique that has the potential to revolutionize agricultural systems [20,26].

1.9 FUTURE DIRECTIONS IN SUSTAINABLE AGRICULTURE

The increasing need for food worldwide, coupled with population increase and climate change, calls for creative ways to boost agricultural output while preserving the environment. The integration of state-of-the-art technologies, resource-efficient methods, and ecologically conscious approaches is fundamental to the future of sustainable agriculture.

A game-changing tool, nanotechnology offers ways to increase crop productivity and nutrient management. An important achievement is the creation of nano-fertilizers, which improve nutrient-use efficiency, lower pollution levels, and facilitate precision farming practices [27]. These developments support environmental sustainability by reducing the misuse of conventional chemical fertilizers and optimizing the delivery of vital nutrients. However, in order to address possible dangers to ecosystems and human health, its implementation necessitates rigorous research and strong regulatory frameworks [28].

Another essential component of future plans is the use of climate-resilient farming techniques. Among these is the creation of agricultural types that are more resilient to pests, drought, and harsh weather. Precision crop alterations made possible by biotechnology advancements like CRISPR-based gene editing promote climate change resilience and adaptability. Furthermore, encouraging integrated farming methods and agroforestry systems helps stabilize ecosystems, diversify agricultural landscapes, and absorb carbon, all of which lessen susceptibilities to the effects of climate change.

It is anticipated that the adoption of precision agriculture technologies would transform farming practices. Farmers may make data-driven decisions by using tools like drones, IoT-enabled sensors, and AI-powered analytics to monitor crop, soil, and water health in real time. More sustainable farming methods are made possible by these technologies, which guarantee the effective use of resources, minimize waste, and increase output [29].

The circular economy concept, which prioritizes the reuse and recycling of agricultural waste, is another essential component of sustainable agriculture. This involves turning animal dung and crop waste into bioenergy and organic fertilizers, which lowers waste and improves soil fertility. These methods support the more general objectives of lowering agriculture's environmental impact and promoting resource efficiency.

In order to advance sustainable agriculture in the coming years, international cooperation and policy assistance will be essential. Farmers will be empowered to shift to more sustainable practices through policies that support the adoption of green technologies, grant access to research funding, and ease the transfer of knowledge. To ensure that these tactics are widely adopted, training initiatives and awareness campaigns can help close the gap between innovation and execution.

The future of agriculture will be determined by the incorporation of cutting-edge technologies, resource-efficient techniques, and climate-smart practices as the world's population continues to grow and the problems posed by climate change continue to exist. These proactive approaches build the groundwork for a robust and sustainable agricultural ecosystem by prioritizing environmental preservation in addition to addressing the urgent need for food security.

1.10 TECHNOLOGICAL ADVANCEMENTS

Sustainable agriculture is about to undergo a revolution because of innovations like biotechnology, artificial intelligence, and precision farming. To maximize resource utilization and raise agricultural yields, precision farming makes use of sensors, GPS, and data analytics. CRISPR gene editing and the creation of genetically modified organisms (GMOs) are two examples of biotechnology that can create crops that are more resilient to pests, illnesses, and harsh weather. In order to overcome the inefficiencies of contemporary agricultural systems, technological innovation is essential, especially when it comes to increasing NUE. We can maximize resource use, boost productivity, and lessen environmental effects by implementing breakthroughs like crop breeding technology, precision farming methods, and sophisticated fertilization techniques, which will help to accelerate the shift to sustainable agriculture [28].

Precision farming makes use of state-of-the-art instruments, such as sensors, data analytics, and satellite-based GPS technology, to facilitate site-specific agricultural resource management. Through real-time data collection on crop health, soil conditions, and weather patterns, precision farming enables farmers to make well-informed decisions and maximize the use of inputs such as pesticides, fertilizers, and water. For example, automated irrigation systems guarantee accurate water distribution, reducing waste, while drones fitted with image sensors can track crop health and pinpoint spots that need attention. This method lessens the environmental impact of farming activities while simultaneously increasing crop yields.

AI has changed the game by providing automation and predictive analytics to solve problems in agriculture. Large-scale datasets are analyzed by machine learning algorithms to anticipate weather, identify pest infestations, and evaluate crop production. For jobs like planting, weeding, and harvesting, AI-powered robots are being utilized more and more, which lowers human costs and boosts productivity. AI-powered decision-support tools assist farmers in implementing best practices suited to particular circumstances, guaranteeing the sustainable management of agricultural resources.

By developing breakthroughs like GMOs and CRISPR-Cas9 gene-editing technologies, biotechnology is essential to the advancement of sustainable agriculture. GMOs are designed to have desired characteristics, like increased nutrient content, drought tolerance, and resistance to diseases and pests. For instance, in many areas, the use of pesticides has decreased due to Bt cotton, which is genetically modified to create a toxin that deters particular pests. Similar to this, CRISPR technology makes it possible to precisely alter crop genomes, leading to the creation of cultivars that are more adapted to shifting climates and can flourish in soils that are low in nutrients. Increased food security, lower input costs, and higher yields are all results of these developments.

One of the biggest inefficiencies in agriculture must be addressed by increasing NUE. Environmental problems, including greenhouse gas emissions and waterway contamination, are caused by the overuse of nitrogen fertilizers. Fertilization practices are changing because of technological advancements, including slow-release fertilizers, microbial inoculants that improve nitrogen uptake, and crops engineered for better NUE. By ensuring that fertilizers are sprayed in the proper amounts and at the right areas, precision application techniques—supported by AI and GPS systems—help to minimize waste and environmental harm.

Sustainable agriculture relies heavily on water management, and intelligent irrigation systems are revolutionizing water use efficiency. Water is only delivered when and where it is needed, thanks to Internet of Things-enabled devices that track soil moisture levels and meteorological data. Water usage is decreased using automated sprinklers and drip irrigation, guaranteeing that crops are properly hydrated without being over-irrigated. These systems are economical and ecologically friendly since they also use less energy.

In order to power farming activities, sustainable agriculture is depending more and more on renewable energy sources like solar, wind, and biomass. For example, solar-powered water pumps offer an affordable option for irrigation in isolated locations. Furthermore, biogas plants provide a circular strategy that minimizes waste and lessens dependency on fossil fuels by turning agricultural waste into energy.

Decision-making in agriculture is changing as a result of the incorporation of digital platforms and big data analytics. Farmers may make strategic decisions that improve sustainability and production by using platforms that provide weather forecasts, market trends, and crop advisory services. By combining data from multiple sources, big data analytics enables thorough analysis and practical insights that enhance farming methods.

1.11 CLIMATE-RESILIENT PRACTICES

Future directions will concentrate on creating and implementing resilience-enhancing strategies because climate change presents serious challenges to agriculture. This entails developing crop types that are adaptable to climate change, putting in place water-efficient irrigation systems, and embracing agroforestry techniques, which incorporate trees and shrubs into agricultural landscapes to improve biodiversity and prevent soil erosion. Understanding and utilizing advantageous symbiotic relationships, like mycorrhiza, offers viable paths toward improving crop output and adaptation in the quest for climate-resilient agricultural methods. Mycorrhizal fungi help plants thrive in a variety of environmental circumstances by encouraging nutrient uptake, water retention, and disease defence. Investigating the molecular mechanisms behind these correlations yields important information for creating climate-smart agriculture practices that are adapted to different climatic conditions and agricultural environments [30].

1.12 SUSTAINABLE SOIL MANAGEMENT

In the 21st century, managing soil sustainably necessitates a comprehensive strategy that tackles a number of issues, such as boosting food production, reducing climate change by sequestering carbon, and improving biodiversity and water usage efficiency [31]. It will be crucial to improve soil health using techniques like organic amendments, cover crops, and no-till farming. Because healthy soils store carbon, hold onto water, and promote strong crop growth, soil management is an important topic of attention. A key component of sustainable soil management is the symbiotic interaction between plants and mycorrhizal fungi, which provides a natural way to improve crop resilience, nutrient cycling, and soil fertility [30].

The complex demands placed on soil resources highlight the necessity of multidisciplinary cooperation between engineers, biologists, economists, social scientists, and soil scientists in order to create novel approaches to sustainable soil management in the face of global issues. At the nexus of environmental stewardship and agricultural sustainability, phosphorus management offers global stakeholders both opportunities and problems. Innovative methods to sustainable management practices are necessary to address the inefficiencies in phosphorus usage, which call for a sophisticated knowledge of stakeholder roles, impact, and transformational potential [32]. Mycorrhizal associations enable sustainable agricultural methods for long-term land stewardship by promoting nutrient uptake and providing pathogen protection, which helps to preserve soil health and productivity.

1.13 CIRCULAR ECONOMY IN AGRICULTURE

The circular economy concept in agriculture aims to maximize resource utilization and minimize waste by emphasizing resource recycling and waste reduction. This entails decreasing food loss along the supply chain, recycling nutrients through livestock manure, and composting organic waste. By reducing dependency on limited resources and reducing the environmental effects of agricultural equipment powered by fossil fuels, the adoption of renewable energy technology in agriculture promotes a circular economy [33].

1.14 POLICY AND INSTITUTIONAL SUPPORT

Promoting sustainable farming practices will require robust institutional frameworks and regulations. This entails giving smallholder farmers easier access to resources and markets, encouraging research and development, and offering incentives for sustainable farming. In order to promote circularity in the agricultural industry and hasten the shift to sustainable farming practices, policy interventions—such as subsidies for farmers to embrace renewable energy—are crucial. In order to advance the concepts of the circular economy, institutional support and cooperative efforts at the local and international levels are essential in promoting the sharing of best practices and information for incorporating renewable energy systems into agricultural operations [5,34].

1.15 GLOBAL COLLABORATION AND KNOWLEDGE SHARING

Increased international cooperation and knowledge sharing will be advantageous for sustainable agriculture in the future. By sharing innovations, research findings, and best practices, international collaboration can guarantee that all regions gain from developments in sustainable agriculture. Promoting knowledge-sharing and technology-transfer programs that aim to increase agricultural sustainability through the broad use of renewable energy solutions requires international cooperation. Industrialized countries might invest in emission reduction initiatives in underdeveloped nations by utilizing clean development procedures. This will promote international cooperation and aid in the wider use of renewable energy systems in agriculture. The goal of sustainable agriculture in the future is to develop resilient, productive, and ecologically friendly food systems that can satisfy the needs of an expanding population while protecting the environment for coming generations.

1.16 CONCLUSION

The long-term sustainability of our food systems and the wellbeing of our planet depend heavily on sustainable agriculture. Conventional farming methods frequently result in increasing greenhouse gas emissions, biodiversity loss, water scarcity, and soil degradation. The ability of future generations to produce enough food is at risk due to these environmental problems. By supporting methods that preserve and improve natural resources, sustainable agriculture seeks to address these issues. Preserving soil health and lowering chemical inputs involve employing strategies like crop rotation, conservation tillage, and integrated pest control. While agroforestry and preserving natural environments promote biodiversity, efficient water usage techniques like drip irrigation preserve valuable water resources.

Furthermore, social justice and economic stability depend on sustainable agriculture. It contributes to the livelihoods of rural communities by guaranteeing that farming continues to be profitable for both large and small farmers. Sustainable agriculture promotes the welfare of all parties involved in the agricultural supply chain by placing a strong emphasis on ethical labour practices and fair access to resources. In the end, implementing sustainable farming methods is crucial to reducing climate change, supplying the world's expanding food needs, and protecting the environment for coming generations. As demonstrated by programs like organic farming and the expanding use of agroecological methods globally, integrating ecological, agronomic, economic, and sociological viewpoints is crucial to attaining agricultural sustainability.

REFERENCES

[1] A. Wezel, M. Casagrande, F. Celette, J. F. Vian, A. Ferrer, and J. Peigné, "Agroecological practices for sustainable agriculture. A review," *Agron. Sustain.* Dev., vol. 34, no. 1, pp. 1–20, 2014, doi: 10.1007/s13593-013-0180-7.

[2] E. Lichtfouse, *Biodiversity, Biofuels, Agroforestry and Conservation Agriculture.* 2010. Springer. doi: 10.1007/978-90-481-9513-8.

[3] V. Piñeiro, J. Arias, J. Dürr, P. Elverdin, A. M. Ibáñez, A. Kinengyere, et al., "A scoping review on incentives for adoption of sustainable agricultural practices and their outcomes," *Nat. Sustain.*, vol. 3, no. 10, pp. 809–820, 2020, doi: 10.1038/s41893-020-00617-y.

[4] M. Agovino, M. Casaccia, M. Ciommi, M. Ferrara, and K. Marchesano, "Agriculture, climate change and sustainability: The case of EU-28," *Ecol.* Indic., vol. 105, no. April, pp. 525–543, 2019, doi: 10.1016/j.ecolind.2018.04.064.

[5] M. C. Appleby, "Sustainable agriculture is humane, humane agriculture is sustainable," *J. Agric. Environ. Ethics*, vol. 18, no. 3, pp. 293–303, 2005, doi: 10.1007/s10806-005-1490-9.

[6] E. P. Svitacova, "Environmental responsibility in sustainable agriculture," *Bulg. J. Agric. Sci.*, vol. 27, no. 5, pp. 880–886, 2021.

[7] X. Zhang, G. Yao, S. Vishwakarma, C. Dalin, A. M. Komarek, D. R. Kanter, et al., "Quantitative assessment of agricultural sustainability reveals divergent priorities among nations," *One Earth*, vol. 4, no. 9, pp. 1262–1277, 2021, doi: 10.1016/j.oneear.2021.08.015.

[8] R. P. Singh, R. Handa, and G. Manchanda, "Nanoparticles in sustainable agriculture: An emerging opportunity," *J. Control.* Release, vol. 329, pp. 1234–1248, 2021, doi: 10.1016/j.jconrel.2020.10.051.

[9] S. Chaudhuri, M. Roy, L. M. McDonald, and Y. Emendack, "Reflections on farmers' social networks: A means for sustainable agricultural development?," *Environ. Dev. Sustain.*, vol. 23, no. 3, pp. 2973–3008, 2021. doi: 10.1007/s10668-020-00762-6.

[10] M. Nematollahi and A. Tajbakhsh, "Past, present, and prospective themes of sustainable agricultural supply chains: A content analysis," *J. Clean.* Prod., vol. 271, p. 122201, 2020, doi: 10.1016/j.jclepro.2020.122201.

[11] J. Streimikis and T. Baležentis, "Agricultural sustainability assessment framework integrating sustainable development goals and interlinked priorities of environmental, climate and agriculture policies," *Sustain.* Dev., vol. 28, no. 6, pp. 1702–1712, 2020, doi: 10.1002/sd.2118.

[12] N. Duić, K. Urbaniec, and D. Huisingh, "Components and structures of the pillars of sustainability," *J. Clean.* Prod., vol. 88, pp. 1–12, 2015, doi: 10.1016/j.jclepro.2014.11.030.

[13] N. Van Cauwenbergh, K. Biala, C. Bielders, V. Brouckaert, L. Franchois, V. Garcia Cidad, et al., "SAFE-A hierarchical framework for assessing the sustainability of agricultural systems," *Agric. Ecosyst. Environ.*, vol. 120, no. 2–4, pp. 229–242, 2007, doi: 10.1016/j.agee.2006.09.006.

[14] M. A. Altieri and C. I. Nicholls, "Agroecology scaling up for food sovereignty and resiliency," *Sustain. Agric. Rev.*, vol. 11, pp. 1–29, 2012, doi: 10.1007/978-94-007-5449-2.

[15] S. Velten, J. Leventon, N. Jager, and J. Newig, "What is sustainable agriculture? A systematic review," *Sustainability*, vol. 7, no. 6, pp. 7833–7865, 2015, doi: 10.3390/su7067833.

[16] B. Purvis, Y. Mao, and D. Robinson, "Three pillars of sustainability: In search of conceptual origins," *Sustain. Sci.*, vol. 14, no. 3, pp. 681–695, 2019, doi: 10.1007/s11625-018-0627-5.

[17] M. A. Altieri, C. I. Nicholls, and R. Montalba, "Technological approaches to sustainable agriculture at a crossroads: An agroecological perspective," *Sustainability*, vol. 9, no. 3, pp. 1–13, 2017, doi: 10.3390/su9030349.

[18] T. Mahanty, S. Bhattacharjee, M. Goswami, P. Bhattacharyya, B. Das, A. Ghosh, and P. Tribedi, "Biofertilizers: A potential approach for sustainable agriculture development," *Environ. Sci. Pollut. Res.*, vol. 24, no. 4, pp. 3315–3335, 2017, doi: 10.1007/s11356-016-8104-0.

[19] M. Farooq and M. Pisante, *Innovations in Sustainable Agriculture.* 2019. Springer. doi: 10.1007/978-3-030-23169-9.

[20] M. Nazirul, I. and Sarker, and M. Wu, "Role of climate smart agriculture in promoting sustainable agriculture : A systematic literature review," *Int. J. Agric. Resour. Gov. Ecol.*, vol. 15, no. 4, pp. 323–337, 2019.

[21] M. Singh and K. L. Maharjan, *Sustainability of Organic Farming in Nepal*, 208 pp. 2017, doi: 10.1007/978-981-10-5619-2.

[22] T. Gomiero, D. Pimentel, and M. G. Paoletti, "Is there a need for a more sustainable agriculture?," *CRC. Crit. Rev. Plant Sci.*, vol. 30, no. 1–2, pp. 6–23, 2011, doi: 10.1080/07352689.2011.553515.

[23] J. Janker, S. Mann, and S. Rist, "Social sustainability in agriculture – A system-based framework," *J. Rural Stud.*, vol. 65, no. December, pp. 32–42, 2019, doi: 10.1016/j.jrurstud.2018.12.010.

[24] A. Gaviglio, M. Bertocchi, M. E. Marescotti, E. Demartini, and A. Pirani, "The social pillar of sustainability: A quantitative approach at the farm level," *Agric. Food Econ.*, vol. 4, no. 1, 2016, doi: 10.1186/s40100-016-0059-4.

[25] E. Gamalero, G. Lingua, G. Berta, and P. Lemanceau. "Methods for studying root colonization by introduced beneficial bacteria". In: E. Lichtfouse, M. Navarrete, P. Debaeke, S. Véronique, C. Alberola (eds) *Sustainable Agriculture.* 2009. Springer. https://doi.org/10.1007/978-90-481-2666-8_37

[26] M. R. Al-Mamun, M. R. Hasan, M. S. Ahommed, M. S. Bacchu, M. R. Ali, and M. Z. H. Khan, "Nanofertilizers towards sustainable agriculture and environment," *Environ. Technol.* Innov., vol. 23, p. 101658, 2021, doi: 10.1016/j.eti.2021.101658.

[27] R. Raliya, V. Saharan, C. Dimkpa, and P. Biswas, "Nanofertilizer for precision and sustainable agriculture: Current state and future perspectives," *J. Agric. Food Chem.*, vol. 66, no. 26, pp. 6487–6503, 2018, doi: 10.1021/acs.jafc.7b02178.

[28] F. Zulfiqar, M. Navarro, M. Ashraf, N. A. Akram, and S. Munné-Bosch, "Nanofertilizer use for sustainable agriculture: Advantages and limitations," *Plant Sci.*, vol. 289, no. August, 2019, doi: 10.1016/j.plantsci.2019.110270.

[29] J. Wu and H. D. Haasis, "The freight village as a pathway to sustainable agricultural products logistics in China," *J. Clean.* Prod., vol. 196, pp. 1227–1238, 2018, doi: 10.1016/j.jclepro.2018.06.077.

[30] S. Basu, R. C. Rabara, and S. Negi, "AMF: The future prospect for sustainable agriculture," *Physiol. Mol. Plant Pathol.*, vol. 102, pp. 36–45, 2018, doi: 10.1016/j.pmpp.2017.11.007.
[31] R. Lal, "Soils and sustainable agriculture. A review," *Agron. Sustain.* Dev., vol. 28, no. 1, pp. 57–64, 2008, https://doi.org/10.1051/agro:2007025.
[32] C. Lyon, D. Cordell, B. Jacobs, J. Martin-Ortega, R. Marshall, M. A. Camargo-Valero, et al., "Five pillars for stakeholder analyses in sustainability transformations: The global case of phosphorus," *Environ. Sci. Policy*, vol. 107, no. February, pp. 80–89, 2020, doi: 10.1016/j.envsci.2020.02.019.
[33] A. Chel and G. Kaushik, "Renewable energy for sustainable agriculture," *Agron. Sustain.* Dev., vol. 31, no. 1, pp. 91–118, 2011, https://doi.org/10.1051/agro/2010029.
[34] M. A. Levy, M. N. Lubell, and N. McRoberts, "The structure of mental models of sustainable agriculture," *Nat. Sustain.*, vol. 1, no. 8, pp. 413–420, 2018, doi: 10.1038/s41893-018-0116-y.

Chapter 2

Emerging technologies and their uses in the agriculture sector

Sonali Kokale, Raj Pooniya, Atin Kumar, Sharad Sachan, and Mohamed A.E. Abdel Rahman

2.1 INTRODUCTION

Agriculture—the bedrock on which civilization rests—must overcome several hurdles in order to face the increasing demands of a burgeoning world population while ensuring that food resources are used sustainably and resiliently. Unfortunately, the traditional methods that have served us well in the past are struggling to keep up with a fast-changing environment and an ever-expanding world population. The agricultural industry passes through a complicated maze of interconnected problems, from the terrible threat of climate change to the continuation of rural poverty.

Emerging technologies have brought revolutionary changes to the agriculture industry. These not only revolutionized the practices of a common type of cultivated agricultural techniques, but also paved the way for new and resilient forms of highly effective and sustainable agricultural systems. This chapter discusses the emerging trends of these new technologies and their relation to the challenges facing agriculture today. It outlines revolutionary solutions that may fundamentally transform the practice of farming in the future. The incorporation of advanced technology has revolutionized modern agriculture and has opened the gates of opportunities for production improvement, maximization of resource use, and resolution of global food security issues. Some examples of such potentials are precision farming and advanced data analytics [1].

This chapter aims to explore how new technology may be harnessed to address problems currently facing the agriculture sector. The primary aim of this chapter is to give a brief overview of the newest emerging technologies that are expected to change the agricultural landscape. It will present concrete cases and application models of technologies such as robotics, blockchain, artificial intelligence (AI), Internet of Things (IoT), and precision agriculture in an attempt to help solve the major issues that farmers and the agriculture sector confront. The scope also includes describing possible challenges, opportunities, and future trends in utilizing and integrating these technologies within agricultural operations, in addition to

 DOI: 10.1201/9781003494782-2

stressing the effects of these technologies on resilience, productivity, and sustainability [2].

Gradually, more and more challenges are facing agriculture and pose a threat to its productivity, sustainability, and ability to feed the ever-growing population of the world. Technological limitations, socio-economic inequalities, and environmental issues are some of these causes. Our capacity to recognize and redress those problems will determine food security, environmental sustainability, and the welfare of agricultural communities around the world.

- **Climate Change:** Hanging heavily over agriculture is the threat of climate change, with its manifestations of erratic weather patterns, extreme events, and shifting climatic zones. Rising temperatures and extended droughts have upset traditional agricultural practices, allowing for the risk of crop production and food security due to erratic rainfall.
- **Resource Scarcity:** Limited supplies of fossil fuels, arable land, and water put further pressure on agricultural systems that can hardly meet rising demand. An increasing demand for water resources from urbanization and industrialization aggravates the water stress in agricultural areas.
- **Loss of Biodiversity:** Biodiversity loss has contributed to intensive agriculture, habitat destruction, and deforestation, ultimately threatening agricultural sustainability and ecological resilience. The decline of pollinator and beneficial insect populations threatens agricultural pollination and pest control services essential for food production [3].
- **Nutrition and Food Security:** In spite of increased production, there are still millions of people who are hungry, malnourished, and food insecure. Persistent poverty, inadequate access to safe food, and inequitable distribution make food security for all hard.
- **Poverty and Inequality Reduction for Rural Communities:** Most of the poverty, marginalization, and struggle around land tenure affect rural communities in the poor property of smallholders in developing countries. Limited access to markets, resources, and financial services reduces their ability to adopt sustainable farming alternatives and improve livelihoods.
- **Diseases and Pests:** A constant danger to crop health and yield is posed by the emergence of new pests and diseases and the spread of existing ones, facilitated by global trade and climate change. Additional challenges to pest management arise from regulations and pesticide resistance.
- **Technological Divide:** New technologies offer huge potential to transform agriculture, but differences in their adoption, accessibility, and digital literacy further worsen inequity. Smallholder farmers are at

risk of being left behind regarding the digital revolution, as they lack the infrastructure and resources to use technology effectively.

- **Carbon Emission:** Anthropogenic activities that lead to the emission of carbon-containing gases into the atmosphere, with the major one being carbon dioxide (CO_2), include the burning of fossil fuels for transport, heating, power, and industrial applications. Cement production, land-use change, and deforestation are another set of major contributors. Accumulation of such emissions in the atmosphere can cause an enhancement of the greenhouse effect, whereby the temperatures of the Earth increase, a phenomenon referred to as "global warming."

It is a worthy resolution to adopt an integrated approach combining technological innovations with sustainability, regulatory reforms, and community participation to solve the described challenges. New technology can help agriculture to move through barriers, build resilience, and enhance a more equitable, sustainable, and food-secure future. AgTech marks a disruptive way of viewing agriculture. The fundamental essence of AgTech is the ability to use data. Through networks of sensors deployed in their fields in conjunction with satellite imaging and advanced data analytics, farmers have been given unique real-time insights into their land and crops. This empowers them to move from a reactive to a proactive mode of operation, where they can maximize yield, eliminate waste, and optimize resource use [4].

2.2 PRECISION AGRICULTURE

Advanced precision farming is the newest type of farming where technology is adapted to improve farm output and conservation of agricultural resources. Farmers can now get their sowing, fertilization, and harvesting done with greater precision because of the use of gadgets like GPS, sensors, drones, and data analytics in agriculture. Using the analysis-driven method allows the farmers to target particular zones within their fields with the agricultural ways and thus achieve more effective use of pesticides and fertilizers as well as water. GPS mapping potential must be referred to precision farming, as this could exhibit more differences in terms of crop vigor, soil composition, and moisture level for farmers compared to manual mapping. It is now possible for farmers to make informed decisions on when and where to irrigate, apply fertilizers, or plant certain crops by using such data to make targeted decisions. In addition, real-time data on crop health could be harvested through sensors deployed across the field, thus making it easier to detect problems such as pests and diseases early and to take preventative measures before the movement. Besides GPS and sensors, drones are becoming another increasingly popular tool for precision agriculture [5].

Cameras and sensors installed in drones will take high-resolution photographs of crops and give information on crop health, growth patterns, and trouble spots to the farmer. Such airborne data will lead to economical and sustainable forms of agriculture since it gives farmers decision-making abilities on where to concentrate their efforts and resources. It could transform the agriculture industry altogether, with the possibility of thousands of percentage increases in productivity while decreasing inputs and eventually feeding a growing population in the world [6].

2.2.1 Technologies related to precision agriculture

Geographic Information System (GIS) and Global Positioning System (GPS): It illustrates here that GPS technology (Figure 2.1) provides farmers with accurate field mapping, tracking of equipment movement, and the possibility for spatial variability analysis of crop performance and soil attributes. Decision-making processes regarding field management and resource allocation can be enhanced through the integration and analysis of spatial data made possible by GIS software (Figure 2.1).

Remote Sensing Technologies: Unmanned aerial vehicles (UAVs), satellites, and ground sensors are typical examples of remote sensors that generate much-needed information about such variables as insect pest infestations, plant health, soil moisture, and other environmental factors. Through these technologies (Figure 2.2), farmers can reduce production loss and efficiently use resources by remotely monitoring field conditions and pinpointing areas needing intervention [7].

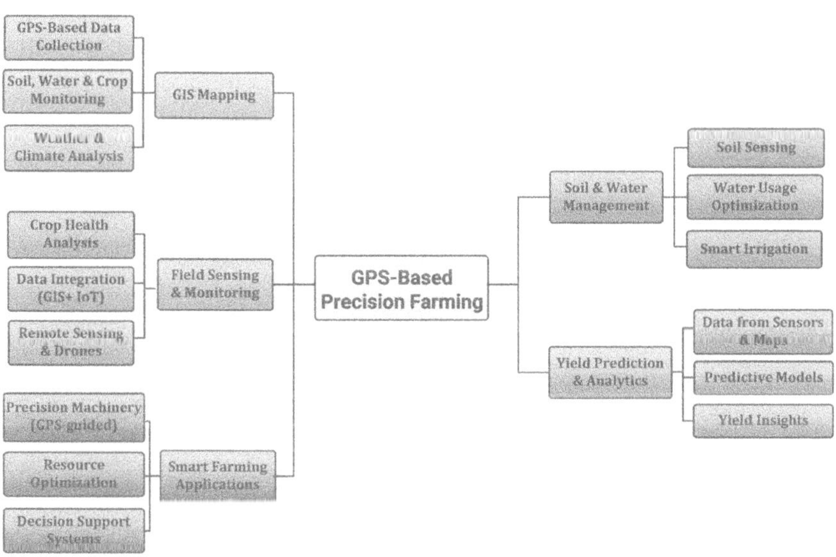

Figure 2.1 Integration of GIS and GPS in agriculture.

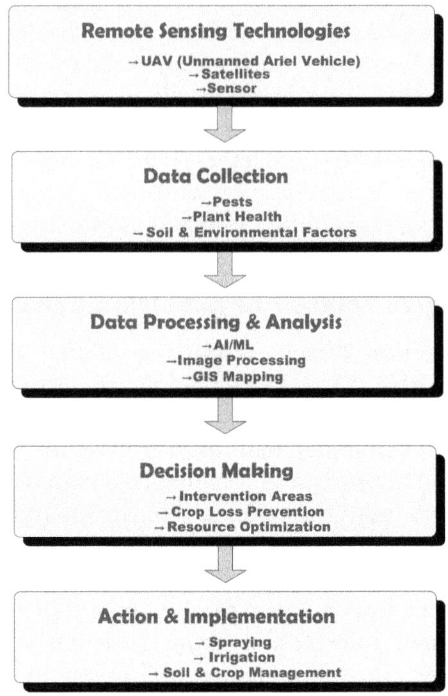

Figure 2.2 Remote sensing technologies in agriculture.

Variable Rate Technology (VRT): Site-specific data and agronomic recommendations form the basis of accurate input application, including fertilizers, herbicides, and irrigation water. VRT (Figure 2.3) reduces runoff of fertilizers and groundwater contamination while producing higher crop yields and improving efficiency in the utilization of resources [8]. This is achieved by tailoring input rates according to the special requirements of different locations in a given field.

Sensor Technology: Advanced sensor technology includes weather stations, sensors for moisture in soil, and sensing devices for crop health. These supply real-time information about the health of crops as well as environmental factors. On the other hand, such timely coordination of interventions in pest control, irrigation timing, and other agronomic practices can improve production and profitability levels for farmers.

Data Analytics and Decision Support Systems (DSS): These are data analytics tools that analyze massive datasets sourced from various inputs such as sensors, drones, and agricultural machines. Together with these data, DSS create useful insights and recommendations for enhancing production methods, forecasting crop yields, and mitigating risks associated with pests, diseases, and adverse weather occurrences for farmers [9].

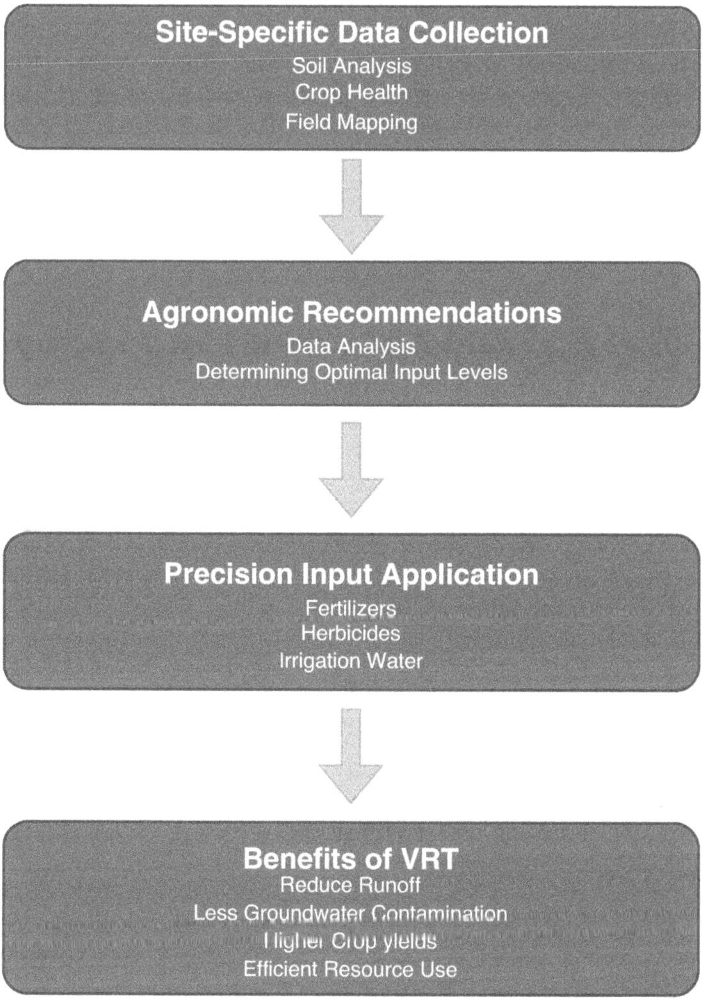

Figure 2.3 Variable rate technology (VRT) in agriculture.

2.2.2 Benefits of precision agriculture

- **Improved Output:** Precision agriculture can yield very high production and be very economically viable by using site-specific variables to customize inputs and management techniques.
- **Resource-Use Efficiency:** Precision agriculture is the application of inputs with increased accuracy, thus reducing waste and reducing the costs of production, keeping natural resources like water and nutrients safe.
- **Environmental Sustainability:** Precision agriculture strengthens sustainable farming practices that create unfavorable environmental effects,

such as soil erosion, water pollution, and greenhouse gas emissions, through focused intervention and optimum resource management.

- **Risk Mitigations:** Precision agriculture lessens susceptibility and increases resilience by helping farmers monitor likely threats, such as pest outbreaks, droughts, and market fluctuations. It does so by equipping them with real-time data in decision support systems.
- **Food Security:** Precision agriculture contributes significantly to ensuring food security by boosting production and enhancing resilience against threats such as population increase and climate change.

2.3 ADVANCED SENSORS AND THEIR APPLICATIONS

Advanced sensors (Figure 2.4) play a vital role in precision farming because they now provide the user with information on crop health and some soil and environmental variables. These sensors can measure many parameters, including temperature, humidity, soil moisture, nutrients, and presence of pests. They are usually pre-integrated into agricultural machinery and linked to the farmer's own data management system for prompt evaluation and decision-making [10].

Advanced sensor applications commonly employed in precision farming include:

- Ground moisture monitoring to ensure proper irrigation techniques are used without wastage of water.
- Monitoring soil nutrient levels to determine precise fertilizer application rates and avoid over-fertilization.
- Early detection of diseases and pests so as to mitigate damage to crops and reduce the need for chemical interventions.
- Crop health monitoring based on indicators like leaf temperature and chlorophyll to avert stress and re-align farming practices, if required.
- Weather monitoring for parameters like humidity, temperature, and wind speed to help with scheduling planting and harvesting dates.

Costly sensors hold great importance in equipping farmers with up-to-date knowledge and information, leading to accurate decision-making, which ultimately means more efficient and sustainable farming practices.

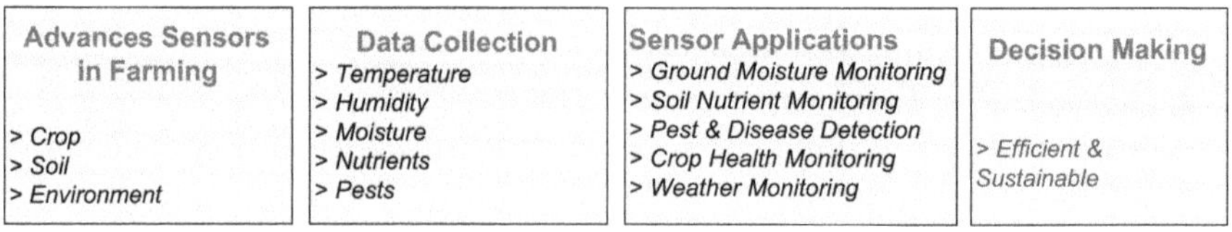

Figure 2.4 Advanced sensors and their applications.

2.4 GPS TECHNOLOGY IN RESOURCE OPTIMIZATION

GPS technology is the backbone of precision farming and resource efficiency. With GPS, farmers can draw accurate maps of their fields with an understanding of variation in the health of crops, moisture content, and soil type. Data-driven approaches help in precise planning by farmers on crop, fertilizer usage, or irrigation timing, thus maximizing production while minimizing costs.

2.5 DATA ANALYTICS FOR IMPROVED CROP YIELD

Data analytics are at the heart of precision farming, whereby enormous data from various sources—sensors, drones, and GPS—are analyzed. Data analytics helps farmers gain valuable insights into crop condition, soil composition, and environmental factors that would enable them to make data-backed decisions for optimizing agricultural yields and resource use. At the University of Illinois, researchers conducted a study highlighting how data analytics assist precision agriculture. It showed how farmers focused their decisions on irrigation and fertilizer applications that would maximize yield and reduce costs based on analysis of data related to crop growth patterns and soil moisture levels [11].

Another study on weather patterns, soil conditions, and past yields has further shown how data analytics can bolster agricultural production estimates. This study, too, was published in Nature Communications. With sophisticated data analytics tools, researchers managed to accurately predict crop yields and suggest ways farmers could boost production through optimal farming practices. To summarize, data analytics is central to the process of precision farming, guiding farmers in making informed decisions that lead to increased crop yields, less resource utilization, and a pathway toward more profitable and sustainable farming systems. Precision agriculture shows promise to deal with the complex issues confronting the global agricultural sector with its transformational farming practices. Precision agriculture employs technology, analytics, and focused management practices to empower farmers with the information and inputs they require to enhance yield in an environmentally friendly manner, conserve resources, and feed a growing population [12].

2.6 AI IN AGRICULTURE

Farmers have now embraced AI, which refers to the enhancement of farmer productivity and a more intelligent choice. By employing technology to collect information on the soil, the weather, and the health of plants, AI helps farmers to grow more food using less water and pesticides. AI allows operation processes to be performed by machinery and robots in tasks such as planting and harvesting for farmers. Farmers save time and money while ensuring environmental protection. AI transforms agriculture and increases productivity. AI is going to revolutionize the agricultural field by offering new ways for the industry to increase productivity, sustainability, and efficiency across all possible farming methods. AI would also enable farmers to face complex agricultural issues, make informed decisions based on data, and minimize the use of resources through sophisticated algorithms and machine learning techniques [13].

2.6.1 Key applications of AI in agriculture

See Figure 2.5.

2.6.1.1 Crop management and monitoring

AI systems are examining sensor data, images from drones, and satellite data in monitoring crop health, pest, disease, and weed identification, and fertilization and irrigation schedule adjustments. With the help of early intervention and accurate managements, these technologies improve yields while lowering input costs.

2.6.1.2 Precision farming

Enormous field maps and variability maps are produced after AI algorithms combine with many sources of information: sensor nets such as GPS and GISs. This helps to maximize use while minimizing environmental effects by giving farmers input on a variable rate, that is, water, fertilizer, and pesticides, in the analysis of spatial patterns and trends across fields.

2.6.1.3 Predictive analytics

Using populations of time series data, AI algorithms would consider past records of agricultural performance, soil properties, market prices, agricultural weather patterns, and so on while predicting pest attacks, crop yield, and market pricing. By relying on predictive insight, farmers can increase income while mitigating risks, formulating plans, and making informed decisions.

2.6.1.4 Robotic automation

Labor-intensive operations such as planting, weeding, harvesting, and sorting are autonomously performed using AI-based robotics and automation technologies. With the combination of robotic systems that use computer vision and machine learning for precise identification and handling, labor costs are saved while boosting the operational efficiency of these systems.

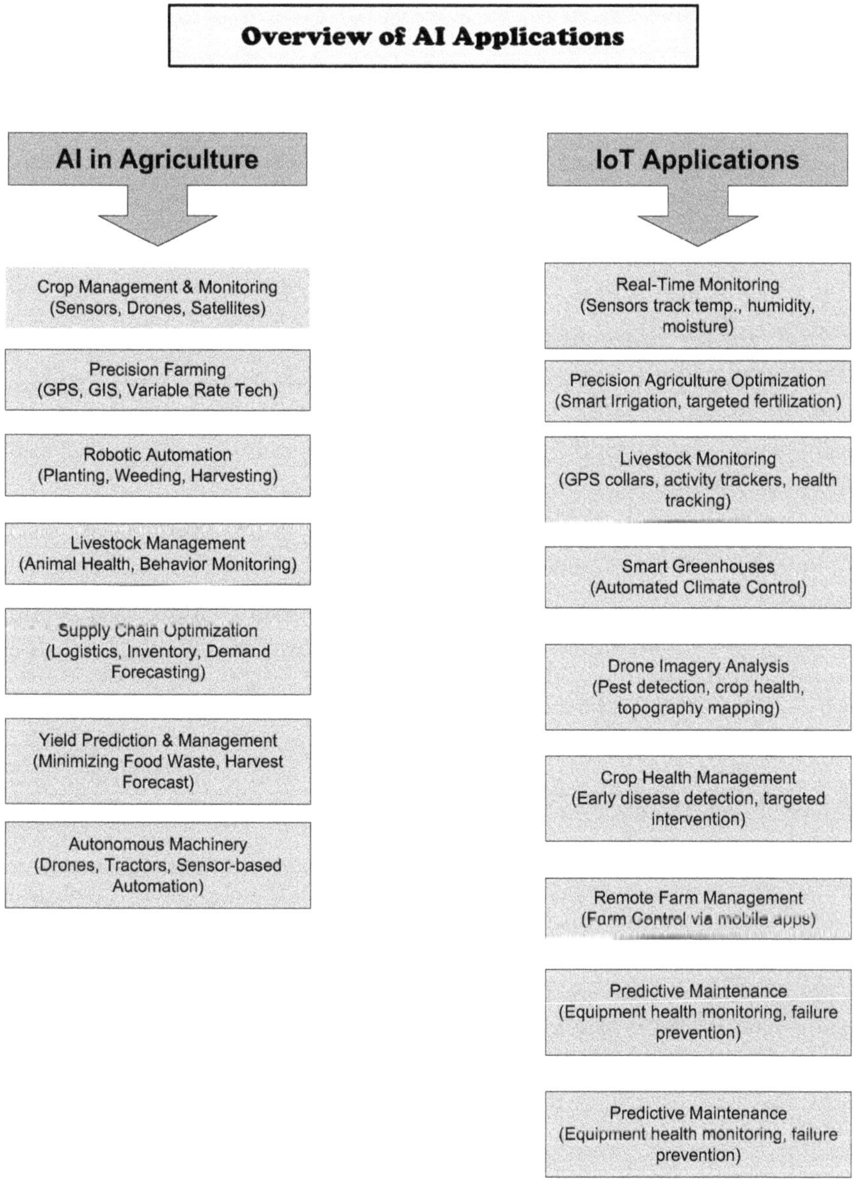

Figure 2.5 Overview of artificial intelligence applications in agriculture.

2.6.1.5 Livestock management and monitoring

In livestock operations, AI-based sensors and monitoring systems monitor environmental variables, animal behavior, and health indicators. Based on real-time analysis of these data, farmers can make adjustments to feeding programs, recognize signs of disease, and thereby improve animal comfort and production.

2.6.1.6 Supply chain optimization

In the agricultural supply chain, AI algorithms maximize distribution networks, inventory control, and logistics. AI helps achieve productivity and sustainability from farm to fork by demand prediction, streamlining transportation routes, and avoiding waste.

2.6.1.7 Supply chain management and yield prediction

To safely forecast the harvest, machine learning algorithms (Figure 2.6) use both forecasted climatic data and the historical yield data. The improvement of channeling food from producers to consumers has the potential to minimize food waste.

2.6.1.8 Autonomous machinery

While planting, weeding, and harvesting, autonomous tractors and drones equipped with sensors and cameras collect crucial data for precision agriculture. These machines eliminate the factor of fatigue, minimize human errors, and permit undertaking large-scale operations.

2.7 IOT IN AGRICULTURE

This IoT is transforming significant disconnected areas, and agriculture is no exception. In fact, IoT is a network of physical things spread everywhere, which are integrated with software, sensors, and communications, thus gathering and sharing information to facilitate intelligent decision making and real-time monitoring. It is providing an uninterrupted stream of information across many dimensions of the farm environment and crop health for farm management in the agriculture sector [14].

How IoT is reshaping the agricultural landscape is discussed in detail below:

2.7.1 Real-time monitoring

Real-time monitoring of field conditions would form the basis of IoT applications in agriculture. A network of sensors is strategically placed throughout the field to collect information such as temperature, humidity, light levels, and even nutrient content. These data can be continuously used by farmers to provide a picture of their fields by transmitting it through wireless communication to a central hub or cloud platform.

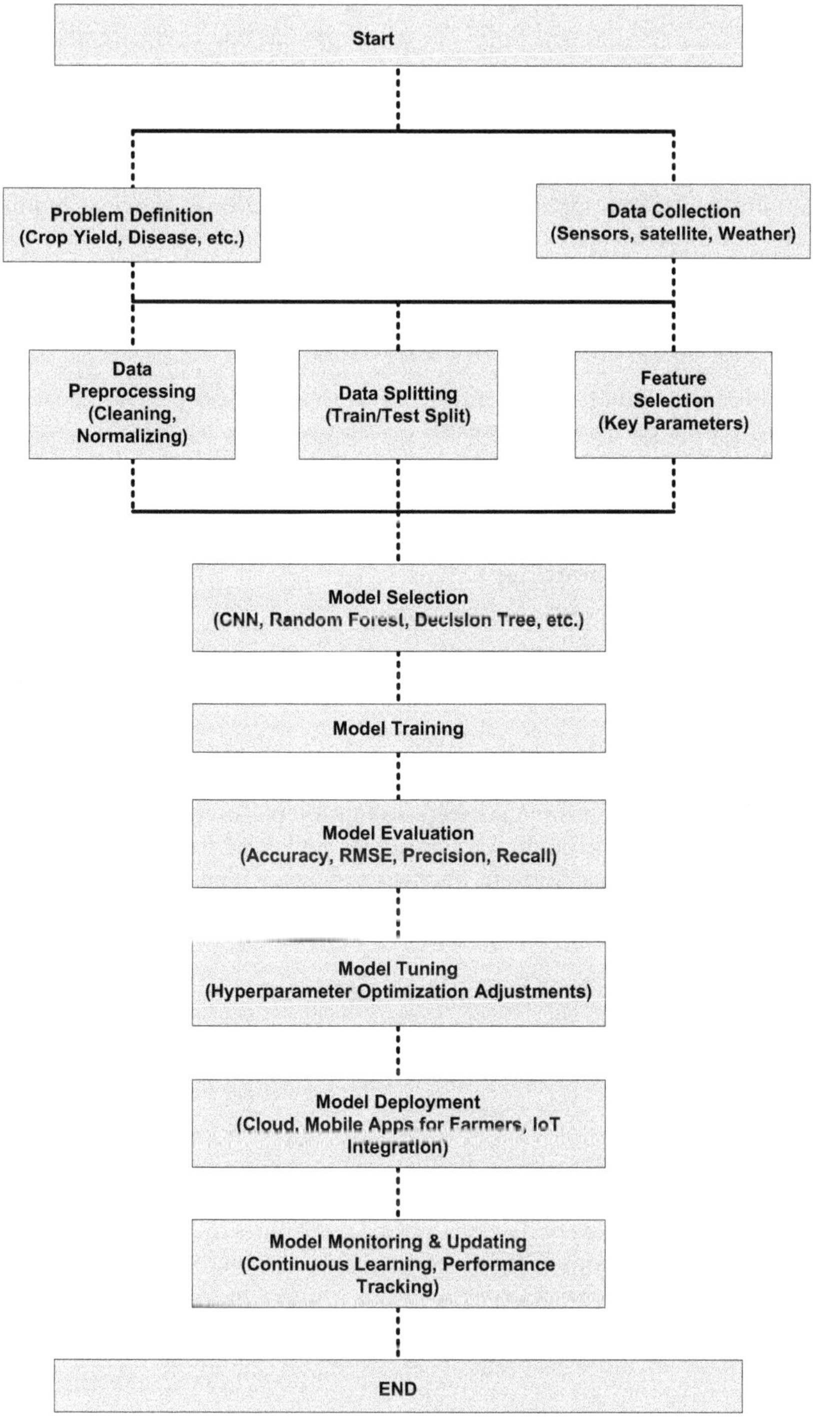

Figure 2.6 Algorithm of machine learning (ML) in agriculture.

2.7.2 Optimization of precision agriculture

This immensely increases the precision of agriculture technologies as real-time data are gathered from the various devices of the IoT. Farmers are now able to use their resources intelligently due to information about field conditions—they can change irrigation schedules if the soil moisture is high, thus guaranteeing not wasting a lot of water. In addition, IoT data may direct fertilizer application according to the specific nutrient requirements in different zones of a field, favorable to proper resource utilization and a reduced impact on the environment.

2.7.3 Livestock monitoring solutions

GPS collars, activity trackers, and wireless weight scales portray the movements, feeding patterns, and health conditions being tracked for animals. This information enables ranchers to quickly attend to the sick animals, maximize their breeding programs, and improve overall wellbeing.

2.7.4 Smart greenhouses

Threshold-based adjustment of temperature, humidity, illumination, and CO_2 levels by climate controllers creates conditions for optimum growth of several plant species.

2.7.5 Drone imagery analysis

The UAV-mounted sensors take high-resolution pictures and videos, which are later analyzed through AI algorithms and lead to framing the field topography, pest identification, nutrient deficiency identification, and crop growth stages.

2.7.6 System for inventory management

It delivers stock data continuously to a central database and connected storage units and silos and cool rooms to prevent loss of stock consistency, spoiling, and theft.

2.7.7 Improved crop health management

Early diagnosis of stress or disease in crops determines the amount of damage and successful harvest. IoT sensors are capable of monitoring certain factors such as temperature variability in leaves or changes in soil moisture content, which may lead to serious problems. Farmers can improve the health and yield of crops by applying targeted pesticides or modifying their irrigation procedure once they are alerted to any emerging menace in good time [15].

2.7.8 Farm management remotely

With IoT, farmers can now monitor and control their farms without being physically present on their farms. Farmers receive real-time information from their farms using mobile apps or web-based dashboards, regardless of where they are. Faster decision-making is enabled, and farm management is more effective, particularly in extensive farms.

2.7.9 Automation and robotics

The IoT is crucial to adopting agricultural robots and automation in farming. Sensors within these robots allow them to gather environmental and crop information and thus improve their accuracy and productivity. An example of this would be autonomous robots with IoT sensors roaming the fields, identifying weeds, and applying targeted herbicide only where necessary to reduce human work and herbicide use. The integration of IoT into agriculture has already shown its benefits, though it is still in its baby stages. Sensor technology will progress; connection prices will drop, and, with that, the agricultural world will experience an even bigger take-up of IoT solutions on farms.

2.7.10 Predictive maintenance

IoT sensors will be used to monitor the states of farming equipment to establish predictive maintenance by detecting possible failures before breakdowns occur. Hence, it can minimize downtime and ensure effective agribusiness operations.

2.7.11 More robust traceability and food security

With IoT tracking agricultural items throughout the distribution supply chain, important data will be available on the origin, storage, and transit of the product. Such transparency can promote food safety and instill customer confidence.

2.7.12 Integration with big data and AI

Combining the real-time data collected from IoT sensors and the analytical power of Big Data and AI will enable deriving far better and deeper insights. AI algorithms now empower agricultural decision making at an unprecedented level by assessing huge amounts of data, spotting trends, being predictive of potential problems, and suggesting optimal management strategies [16].

2.8 BLOCKCHAIN TECHNOLOGY IN AGRICULTURE

Blockchain (Figure 2.7) is one technology that has garnered the interest of sectors, while its decentralized, transparent, and secure nature would benefit agriculture. Blockchain essentially provides verifiable proof of origin, legitimacy, possession, and transfer of the commodities by recording transactions in a distributed ledger. It thus efficiently aids in fraud prevention, supply chain streamlining, and consumer confidence building. This section treats the features, advantages, drawbacks, and glaring instances of blockchain technology in agriculture.

2.8.1 Functionalities

Digitized Identification: Supply chains may be completely tracked, traced, and audited by embedding cryptographically secure IDs onto goods, packages, or shipments.

Smart Contracts: Programmatically encoded contracts that automatically execute certain activities without any manual intervention to encourage autonomy once predetermined criteria have been satisfied.

Tokenization: The trade of tangible assets represented as fungible tokens then lays the ground for trade discussions, escrow agreements, and payment settlements, thereby stimulating trade.

Consensus Mechanisms: They eliminate single points of failure, hostile assaults, and counterfeiting attempts by enabling network members to jointly validate transactions.

2.8.2 Benefits

Increased Trust: The validation of provenance along agri-food supply chains strengthens the producer-consumer relationship by guaranteeing that agri-food supply is legitimate, dependable, and of good integrity.

Increased Transparency: Publication of transactional facts promotes accountability, transparency, and honesty; this discourages unethical practices.

Enhanced Traceability: Accurate documentation supports rapid response to food safety threats, thus mitigating financial losses and reputational damage.

Simplified Trade: Automated execution, uncomplicating payments and standardized contract templates, eliminating paper trails, red tape, and bureaucratic bottlenecks.

2.8.3 Drawbacks

There are lots of attractive sides to blockchain, but there are still a number of things that can be rightfully labeled as points against it—scalability concerns, energy consumption, regulatory uncertainty, and user-unfriendliness.

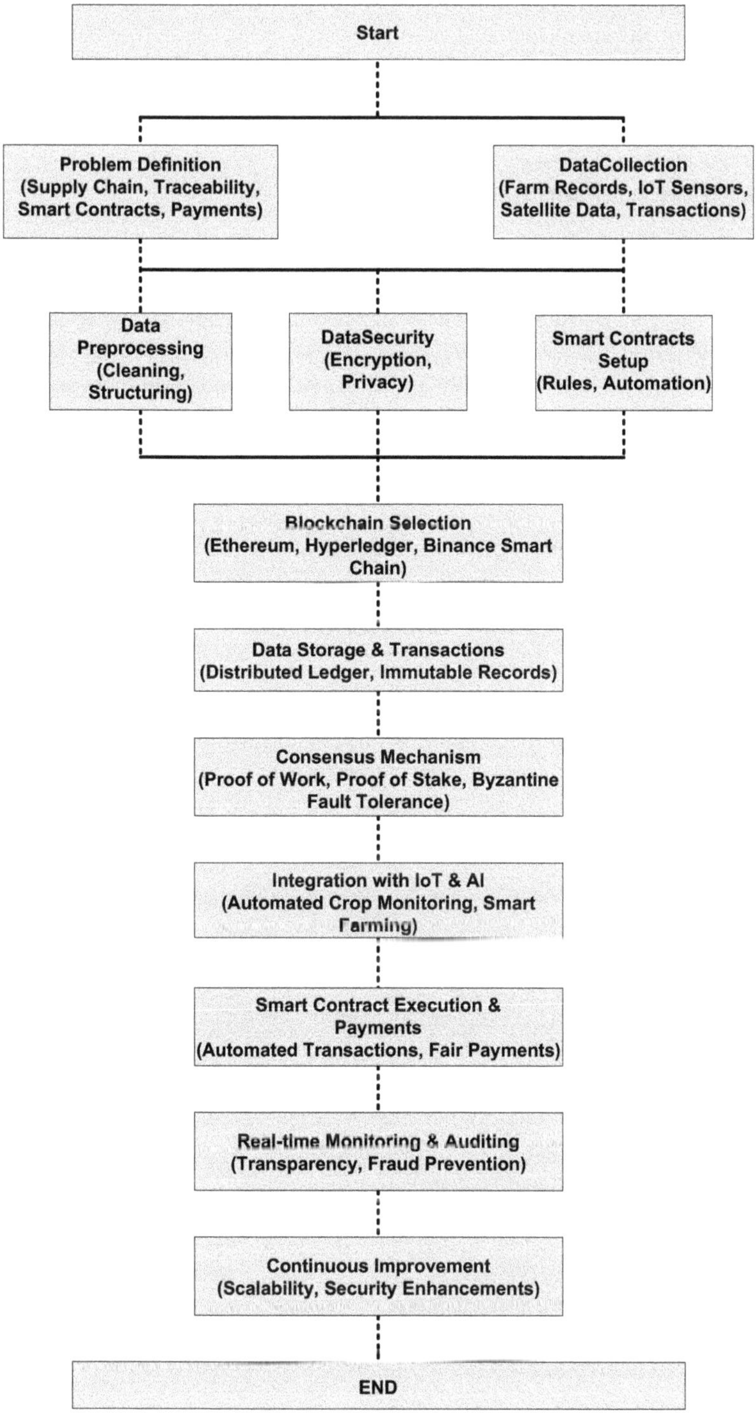

Figure 2.7 Blockchain technology in agriculture.

Initial decisions can also be badly affected by things such as steep learning curves, a big initial input, and cloudy return on investment. Yet when the progress toward solutions to the problems continues, the road shall be paved for general acceptance.

2.9 VERTICAL FARMING

The unconventional and sustainable agricultural practice called vertical farming (Figure 2.8) cultivates crops and grows them in vertical, stacked tiered systems, usually within a building or an enclosed structure. It represents one break from the oldest of traditional agricultural practices: the almost uniformly defined horizontal agricultural practice of growing crops in open fields, the others being conventional farming per se and organic agriculture. The vertical farming system is envisioned as a solution to critical problems afflicting modern agriculture—these problems include unavailability of land, unavailability of water, environmental degradation, and climate change [17]. Various components of vertical farming are given in Table 2.1.

2.9.1 Some of the major characteristics of vertical farming

Manipulation of growth parameters such as CO_2 concentration, temperature, humidity, and light intensities is very carefully performed and kept within ideal limits to ensure constant output even with changing weather conditions outside. Plants are cultivated in stacked layers to maximize land use efficiency for intensive cultivation. Examples range from simple shelving

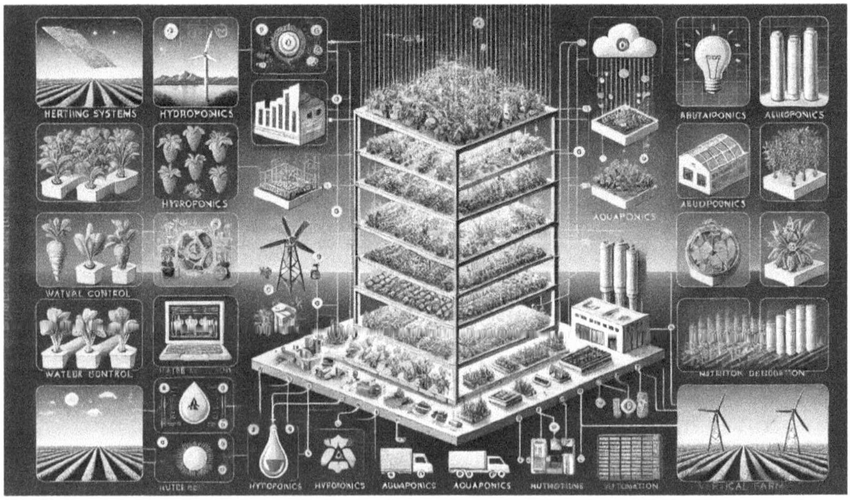

Figure 2.8 Smart vertical farming components.

Table 2.1 Various components of vertical farming

S. No.	*Component*	*Description*
1	Growing System	Hydroponics, Aeroponics, Aquaponics
2	Lighting	LED Grow Lights, Artificial Sunlight
3	Climate Control	Temperature, Humidity, CO_2 Levels
4	Water Management	Drip Irrigation, Water Recycling
5	Nutrient Delivery	Automated Fertilizer Dosing
6	Automation IoT	Smart Sensors, AI-based Monitoring
7	Energy Sources	Solar, Wind, Grid-based Power
8	Structure	Multi-layer Racks, Vertical Towers
9	Crop Selection	Leafy Greens, Herbs, Strawberries, Microgreens
10	Blockchain Integration	Supply Chain Transparency, Smart Contracts
11	Data Analytics	Yield Prediction, Disease Detection
12	Waste Management	Composting, Circular Economy

systems to very complex structures and multi-story building-type systems. Nutrient-dissolved water runs over root systems, giving vital minerals for fast development. Water wastes are kept minimal, with the recirculation of any runoff after filtration and replenishment. The spectral emission of specialized light-emitting diode (LED) lighting systems is tailored to the photosynthetic requirements of plants, enhancing spectral uniformity and energy conversion efficiency. Advanced sensors and actuators enable precise control of growth parameters and automated operation, minimizing human involvement and labor cost [18].

2.9.2 Advantages of vertical farming

Indoor agriculture shields crops from seasonal variations so that a steady yield is achieved in almost any weather. The enclosed environments allow for rational use and recirculation of resources, especially water and nutrients. Localized production minimizes the greenhouse gas emissions from production processes, land disruption, and transportation distance. Growing crops away from outdoor pests and pathogens minimizes the chance of outbreaks of diseases and reduces the need for chemical fungicides and pesticides. Vertical farming helps cultivators grow produce of a variety of colors, sizes, and cultivation periods, improving the variety available for selling or eating [19].

2.9.3 Some drawbacks to vertical farming

Often, high recurrent energy costs and capital expenditures limit entry and growth. Production scaling is partly constrained by physical dimensions, though this constraint is somewhat alleviated through modular design.

The relative newness implies a lack of past research and experience, which makes trial-and-error in creation a necessity. Perceptions of vertical farming based on unfounded beliefs obstruct the work of creating advocates for change and formal policy [20].

2.10 CASE STUDIES

New technologies are beginning to revolutionize the agricultural industry. Their focus is primarily on raising levels of profit, sustainability, and productivity. The following three are case studies in real-life applications of advanced agricultural innovations:

2.10.1 AUGEIAS case study

An intelligent IoT management platform for the reuse of treated wastewater in precision agriculture, the AUGEIAS project enables the reuse of wastewater. It seeks to form an ecosystem that enhances crop production efficiency and minimizes environmental conditions from agricultural practices that have environmental impacts on the use of chemicals and freshwater. Advanced IoT technologies, cloud computing, data analytics, and AI are used by the system to help farmers and water utility companies in decision making [21].

In the north-west region of Greece, a pilot project was divided into three sections: one irrigated with treated wastewater coming from a nearby wastewater treatment plant, one irrigated with common water, and the third a rainfed section. On the other hand, each of these areas is supplied with sensors that monitor different parameters in real time (weather conditions, soil health, and quality of crop).

It uses the LoRaWAN protocol for data transmission and suggests a hybrid protocol optimized for energy-autonomous, cable-less, multi-gateway LoRa networks, as this is efficient for harsh environments, where electricity supply may be unreliable. The system brings the data from multiple sources, that is, IoT sensor data, open data, and laboratory tests, and assesses their reliability [22].

A part of the AUGEIAS smart ecosystem would involve smart applications that help in enhancing irrigation efficiency. It automates irrigation processes and works daily, comparing real-time soil moisture levels to predefined thresholds, using one of the possible mechanisms. It provides a mechanism to accurately determine the amount of water to be irrigated, which helps in water conservation practices.

Additionally, the project investigates the utilization of blockchain technology to boost the water quality data security and transparency by recording all the water quality-related transactions on an immutable ledger. It also integrated edge computing to enable faster data processing and make decisions at the network's edge.

AUGEIAS initiative goes beyond that of claiming the success of technology usefulness in optimizing irrigation management, to the use of machine learning techniques to predict crop production outcomes under the forecast of the weather, enabling improved water resource management, increased crop yield, and decreased environmental impact [23].

2.10.2 VELOS case study

VELOS is a smart ecosystem, an integrated pesticide management for bean cultivations, based on intelligent systems. In the process, based on IoT, UAVs, unmanned ground vehicles (UGVs), and AI, it builds the tools for farmers to make informed decisions on pesticide application in precision agriculture.

Four pilot fields in the Prespa field of Greece are operationally incorporated into a cohesive assembly of interrelated subsystems that deterministically evaluate pest infestation risk and predict pesticide application. One key part of the VELOS System consists of wireless sensors that acquire real-time data from the LoRaWAN protocol, as well as open data from the VELOS trusted external open data [24].

Under a "UAV model as a service" framework, the UAV subsystem of VELOS operates in an efficient way: it manages energy as required by each farmer's requests. Autonomous robotic mechanisms are used by the UGV subsystem to improve data quality for farmers to make accurate predictions about possible pest threats and diseases. Integration of these results in a better pesticide application with timely and efficient operations.

VELOS overscores the importance of the user interface, which makes it easier for farmers to use IoT technologies. The project studies conversational user interfaces, where AI-based chatbots, for example, are used to reduce interactions, with the system reducing the barrier for those who need to use the system. It points out the concept of multimodal communication that collects human–machine interaction to coordinate different agricultural operations.

VELOS provides evidence of successful utilizations of emerging technologies aimed at solving specific agricultural challenges. VELOS seeks to minimize pesticide use as well as improve crop health and increase farmers' productivity by optimizing pesticide usage through intelligent scheduling and real-time data assessment. Innovative about the project not only advances precision agriculture, but the project itself is a model for future smart farming initiatives.

2.11 FUTURE TRENDS AND DEVELOPMENTS

More and more intelligent robots will do difficult tasks such as fruit picking, milking cows, and tilling soil by themselves as robotics and automation strive to move onward. In addition to changes in crop types that will be very much factored in by CRISPR-Cas9, gene editing technologies are expected to improve the types of crops considerably while also enhancing overall productivity, strengthening tolerances to either abiotic stresses or diseases. Controlled-environment agriculture—the growing of fresh vegetables and herbs in stacked levels of growth surfaces—will be very much expanded in urban areas, where there's proximity to a metropolitan area. Scientists would like to establish self-sufficient closed-loop life support systems in homes and recycle organic waste to grow food using advanced hydroponics and aeroponics as the speed of travel into space increases.

2.12 MAJOR CURRENT RESEARCH IN AGRICULTURAL TECHNOLOGIES

Continuous research efforts focused on developing resource integration, such as satellites, drones, sensors, and other connected devices, and extracting useful information for immediate decision-making. Researchers are creating virtual simulations of entire agricultural ecosystems, including interactions of crops with soil, water, and atmospheric factors, to optimize resources and simulate climate impacts on them. Quantum computing has the ability to process complex calculations hundreds of times faster than a conventional computer, which will revolutionize industries such as material science, protein folding, and genomics. Research is being carried out to identify specific bioenergy feedstock with high yields, low input requirements, and beneficial carbon sequestration features in order to renew chemicals and fuels while conserving the habitat of native ecosystems and crops.

2.13 OUTCOME PREDICTIONS FROM SOME OF THESE TRENDS AND INNOVATIONS IN THE AGRICULTURAL SYSTEMS

Agriculture has much to gain from these future innovations and changes that affect resilience, sustainability, and productivity impacts on the agriculture sector. They present novel ways of addressing perennial problems facing farmers like erratic weather patterns, limited resources, and very unpredictable markets, while opening up new opportunities for creativity. The agricultural community may thoroughly prepare itself to welcome new innovations while translating their potential into collective benefit by being abreast of current research and breakthroughs.

2.14 CONCLUSION

Thus, new technologies are having a very promising impact on the agricultural sphere, showing unprecedented increases in profitability, sustainability, and productivity. From precision agriculture and drones doing aerial surveys to AI for automated decision-making and blockchains for supply network protection, these are just some examples, and this is only the beginning. Because of their strong capability to build improved farmer competencies, effective resource use, and insightful data, they are considered the next generation for the future of agriculture. Although the evidence is strong, much work is still ahead. Research into agricultural technologies such as digital twins, quantum computing, and space agriculture is still ongoing and has the prospect of significantly affecting the future of farming. Emerging features such as electronic harvesting, microbial fertilizers, and living construction materials would give rise to an interesting, albeit early-in-their-development future for agriculture. To fully harness the promise of these new innovations and technologies, the agricultural community must continually engage in research and collaboration for technological integration. The keystone actions required for effective navigation through the changing world of technology are the sharing of best practices, pooling of resources, and promotion of dialogue among institutions, governments, and private businesses. It is only then that society would be able to harness the emerging technology in the fullest sense to make mainstream what once were impossible goals for future generations.

REFERENCES

1. M. O. Vaintrub, H. Levit, M. Chincarini, I. Fusaro, M. Giammarco, and G. Vignola, "Review: Precision livestock farming, automats and new technologies: Possible applications in extensive dairy sheep farming," *Animal*, vol. 15, no. 3, p. 100143, Dec. 2020, doi: 10.1016/j.animal.2020.100143.
2. N. Khan, R. L. Ray, G. R. Sargani, M. Ihtisham, M. Khayyam, and S. Ismail, "Current progress and future prospects of agriculture technology: Gateway to sustainable agriculture," *Sustainability*, vol. 13, no. 9, p. 4883, Apr. 2021, doi: 10.3390/su13094883.
3. A. Al-Fuqaha, M. Guizani, M. Mohammadi, M. Aledhari and M. Ayyash, "Internet of things: A survey on enabling technologies, protocols, and applications," *IEEE Communications Surveys & Tutorials*, vol. 17, no. 4, pp. 2347–2376, Fourthquarter 2015, doi: 10.1109/COMST.2015.2444095
4. E. M. B. M. Karunathilake, A. T. Le, S. Heo, Y. S. Chung, and S. Mansoor, "The path to smart farming: Innovations and opportunities in precision agriculture," *Agriculture*, vol. 13, no. 8, p. 1593, Aug. 2023, doi: 10.3390/agriculture13081593.
5. M. Attaran and B. G. Celik, "Digital twin: Benefits, use cases, challenges, and opportunities," *Decision Analytics Journal*, vol. 6, p. 100165, Jan. 2023, doi: 10.1016/j.dajour.2023.100165.
6. M. Javaid, A. Haleem, I. H. Khan, and R. Suman, "Understanding the potential applications of artificial intelligence in agriculture sector," *Advanced Agrochem*, vol. 2, no. 1, pp. 15–30, Oct. 2022, doi: 10.1016/j.aac.2022.10.001.

7. O. Friha, M. A. Ferrag, L. Shu, L. Maglaras, and X. Wang, "Internet of things for the future of smart agriculture: A comprehensive survey of emerging technologies," *IEEE/CAA Journal of Automatica Sinica*, vol. 8, no. 4, pp. 718–752, Mar. 2021, doi: 10.1109/jas.2021.1003925.
8. O. Maghazei, M. A. Lewis, and T. H. Netland, "Emerging technologies and the use case: A multi-year study of drone adoption," *Journal of Operations Management*, vol. 68, no. 6–7, pp. 560–591, Jun. 2022, doi: 10.1002/joom.1196.
9. S. A. Osinga, D. Paudel, S. A. Mouzakitis, and I. N. Athanasiadis, "Big data in agriculture: Between opportunity and solution," *Agricultural Systems*, vol. 195, p. 103298, Oct. 2021, doi: 10.1016/j.agsy.2021.103298.
10. S. O. Araújo, R. S. Peres, J. Barata, F. Lidon, and J. C. Ramalho, "Characterising the agriculture 4.0 landscape—Emerging trends, challenges and opportunities," *Agronomy*, vol. 11, no. 4, p. 667, Apr. 2021, doi: 10.3390/agronomy11040667.
11. S. Rolandi, G. Brunori, M. Bacco, and I. Scotti, "The digitalization of agriculture and rural areas: Towards a taxonomy of the impacts," *Sustainability*, vol. 13, no. 9, p. 5172, May 2021, doi: 10.3390/su13095172.
12. R. Sparrow and M. Howard, "Robots in agriculture: Prospects, impacts, ethics, and policy," *Precision Agriculture*, vol. 22, no. 3, pp. 818–833, Oct. 2020, doi: 10.1007/s11119-020-09757-9.
13. E. Elbasi, N. Mostafa, Z. AlArnaout, A.I. Zreikat, E. Cina, and G. Varghese "Artificial intelligence technology in the agricultural sector: A systematic literature review," *IEEE Access*, vol. 11, pp. 171–202, Dec. 2022, doi: 10.1109/access.2022.3232485.
14. S. Gorjian, O. Fakhraei, A. Gorjian, A. Sharafkhani, and A. Aziznejad, "Sustainable food and agriculture: Employment of renewable energy technologies," *Current Robotics Reports*, vol. 3, no. 3, pp. 153–163, May 2022, doi: 10.1007/s43154-022-00080-x.
15. P. Singh and A. Kaur, "A systematic review of artificial intelligence in agriculture," in: *Deep Learning for Sustainable Agriculture*, Elsevier eBooks, pp. 57–80, Jan. 2022, doi: 10.1016/B978-0-323-85214-2.00011-2
16. A. Rehman, T. Saba, M. Kashif, S. M. Fati, S. A. Bahaj, and H. Chaudhry, "A revisit of Internet of things technologies for monitoring and control strategies in smart agriculture," *Agronomy*, vol. 12, no. 1, p. 127, Jan. 2022, doi: 10.3390/agronomy12010127.
17. S. Menon and K. Jain, "Blockchain technology for transparency in agri-food supply chain: use cases, limitations, and future directions," *IEEE Transactions on Engineering Management*, vol. 71, pp. 106–120, 2024, doi: 10.1109/TEM.2021.3110903.
18. R. Abbasi, P. Martinez, and R. Ahmad, "The digitization of agricultural industry – A systematic literature review on agriculture 4.0," *Smart Agricultural Technology*, vol. 2, p. 100042, Feb. 2022, doi: 10.1016/j.atech.2022.100042.
19. R. K. Singh, R. Berkvens and M. Weyn, "AgriFusion: An architecture for IoT and emerging technologies based on a precision agriculture survey," *IEEE Access*, vol. 9, pp. 136253–136283, 2021, doi: 10.1109/ACCESS.2021.3116814
20. S. A. Bhat and N.-F. Huang, "Big data and AI revolution in precision agriculture: survey and challenges," *IEEE Access*, vol. 9, pp. 110209–110222, 2021, doi: 10.1109/ACCESS.2021.3102227

21. A. Valkanis, G. A. Beletsioti, K. Kantelis, P. Nicopolitidis, G. Papadimitriou, and M. Louta, "Developing energy autonomous and cable-less multi-gateway LoRa networks," in *Proc. Int. Conf. Commun., Comput., Cybersecurity, Informat. (CCCI)*, Chongqing, China, Oct. 2023, pp. 1–8, doi: 10.1109/ccci58712.2023.10290141.
22. Y. J. N. Kumar, V. Spandana, V. S. Vaishnavi, K. Neha, and V. G. R. R. Devi, "Supervised machine learning approach for crop yield prediction in agriculture sector," in *Proc. 5th Int. Conf. Commun. Electron. Syst.*, Coimbatore, India, Jun. 2020, pp. 736–741.
23. K. Banti, I. Karampelia, E. Tsipis, T. Dimakis, N. Papadimitriou, A.-A. A. Boulogeorgos, T. Kyriakidis, T. Karamitsou, C. Karapiperis, S. Lappos, E. Bagkavou, T. Adamidis, I. Gkanatsa, and M. Louta, "Intelligent mechanisms for irrigation optimization via treated wastewater management in precision agriculture—The AUGEIAS example," *Environmental Sciences Proceedings*, vol. 2022, no. 1, p. 50, Oct. 2022.
24. I. Karampelia, T. Kyriakidis, and M. Louta, "UAV swarms & task allocation: The way ahead in precision agriculture," in *Proc. 14th Int. Conf. Inf., Intell., Syst. Appl. (IISA)*, Volos, Greece, Jul. 2023, pp. 1–63.

Chapter 3

Investigating the role of artificial intelligence for ethical and sustainable food production

Nishant Kumar Pathak, Vishal Jayaswal, Sharma Ji, Pushkal Kumar Shukla, and Aditya Pratap Singh

3.1 INTRODUCTION

The world population's explosive growth means there has never been such an imperative need to manufacture ethical and biological foods [1]. Feeding billions is becoming a logistical nightmare, and the necessity to protect the environment, advance animal welfare, and guarantee fair distribution of food is all increasingly pressing on traditional farming techniques. Allow me to introduce artificial intelligence (AI), a tech miracle that could revolutionize the agri-industry. This introduction examines the potential benefits, challenges, and implications of integrating AI technology into agricultural workflows [2] concerning ethical and sustainable food production. In the global food chain, there is a moment of turn. There is a pivot point in the world food web. According to the UN, there will be approximately 10 billion people living on the planet by 2050, so food production would have to grow some 70% to satisfy demand. However, neither social justice nor environmental health needs be compromised in the name of intensification [3]. You start with modern agricultural methods and the fact that it's stripping the land of soil and creating deforestation and depleting freshwater that contribute to climate change

Moreover, moral issues—labor practices, the convenience of food, and animal welfare—show just how badly agriculture must change. Sustainable food production aims to find a balance between social responsibility, environmental stewardship, and the need to produce more food. This means establishing policies that promote responsible working practices, preserve biodiversity, and reduce carbon footprints. Ethical factors also emphasize humane care of animals, transparency in the origin of food, and equity in food access as a means to alleviate food insecurity. The transition to AI in food production presents challenges of significant ethical importance despite the many opportunities it provides [4]. The proper means of implementing AI in the agricultural and food systems are only emerging, and they must be done so through the lens of ethical considerations that ensure emerging technology promotes the well-being of society. Important ethical considerations include data privacy, service impact, and fair sharing of

 DOI: 10.1201/9781003494782-3

rewards. Since AI depends upon huge amounts of data for processes, data privacy is a major issue. Farmers and some customers must trust that these will sequentially be used responsibly and protected from unauthorized use. In addition, the use of AI in food production can lead to the loss of jobs in rural places where employment is already limited [5]. Therefore, it's not just about spreading AI in all domains, we need to be ensure that all stakeholders across the agricultural production ecosystem must benefit from AI, including growers who currently lack access to or cannot afford the latest technologies. Promoting a fairer and more sustainable food system requires providing equal access to AI tools, resources, and training for small farms, large farmers, and agribusinesses alike.

The year-wise AI impact on sustainable food production is shown through a bar graph in Figure 3.1, which shows the increasing impact of AI on ethical and sustainable food production from 2002 to 2024. Over the years, AI's role has grown significantly, as evidenced by the steady rise in the hypothetical contribution level. Key advancements include enhanced precision agriculture, better food supply chain management, and improved waste reduction, reflecting AI's evolving importance in promoting sustainable practices in the food industry. The smart agricultural ecosystem depicts the role and contribution of various technologies, including AI, IoT, drones, and satellites, which work together to ensure the positive and sustainable production of food. Weather patterns, soil conditions, and crop health can all be determined from satellite images, making precise decision-making possible. It consists of a system of surveillance drones that monitor both fields and livestock, detecting issues such as crop diseases or animals in distress so that the problem can be dealt with as rapidly as possible. AI algorithms utilize real-time data collected by IoT sensors from soil moisture, temperature, and nutrient

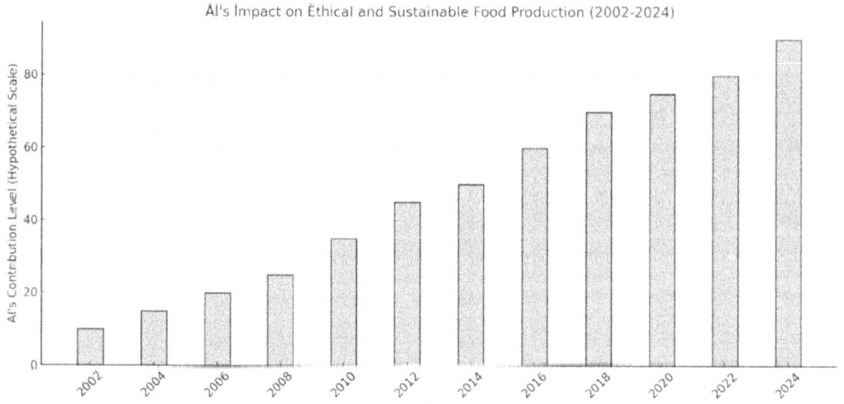

Figure 3.1 Year-wise AI impact.

levels to systemically optimize irrigation and fertilization practices to avoid waste and mitigate environmental damage. One such approach is AI-powered precision agriculture, exemplified by automated machinery allowing water and pesticides, and so on, to be applied only where and when needed, which in turn reduces over-application. Renewable energy wind turbines represent the modernization of agriculture with sustainable power solutions to lower its carbon footprint. Livestock tracking encourages ethical practice through sensors and drones tracking the health and behavior of animals. Farmers can access this networked system through mobile phones, using AI-based tools to manage farms effectively, encouraging ethical practice and maximum resource utilization. This integration shows how technology can address food security challenges while being environmentally and ethically sustainable. In view of the purpose of AI for ethical and sustainable food production, it is important for several reasons.

Dealing with Global Issues: Food Demand and Population Growth: As the world's population increases [6], so does the demand for food. Figure 3.2 shows the smart technologies used in precision agriculture. Maximizing food output to sustainably fulfill this need could make AI quite crucial. Traditional agriculture greatly increases water use, deforestation, greenhouse gas emissions, and other aspects [7]. By means of improved

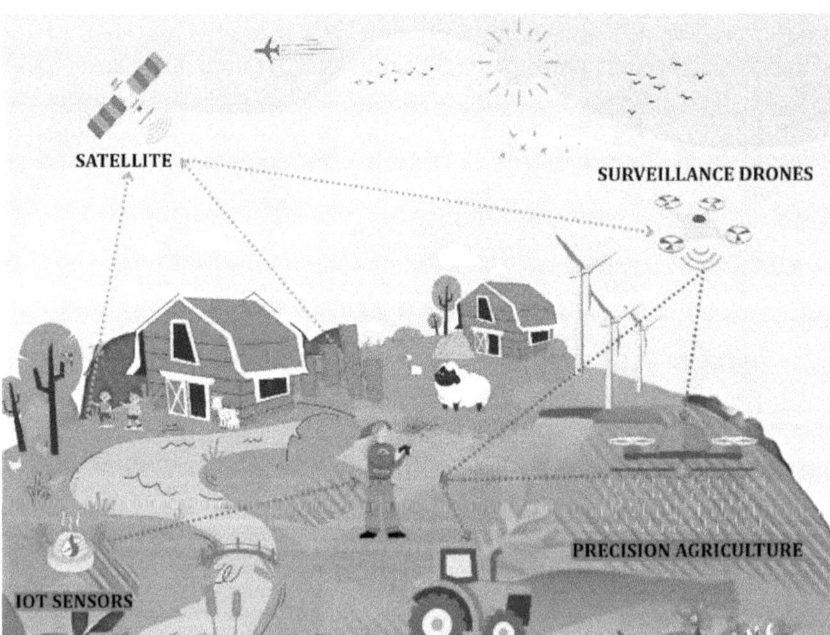

Figure 3.2 Smart technologies used in precision agriculture.

resource management and higher production, AI-driven solutions can help to mitigate these environmental impacts.

- Using data and algorithms, AI can help to improve precision agriculture methods, thereby optimizing planting, fertilizing, and irrigation. From this comes lower resource waste and higher crop yields. AI can streamline the whole food supply chain, from production to distribution, therefore lowering food waste and raising the efficiency of transporting food from farm to table.
- AI can monitor and improve the conditions under which animals are maintained, therefore assuring moral treatment and boosting the standard for animal welfare [8]. AI can help to assure fair working conditions in the agriculture sector, automate labor-intensive activities, and help to minimize labor exploitation.
- **Advancing Sustainable Practices Resource Management:** By optimizing the use of resources like water, fertilizer, and pesticides, AI can lead to more ecologically friendly farming practices that protect ecosystems. Climate Change Adaptation: AI can assist farmers in adapting to climate change by providing data-driven insights on weather patterns, soil conditions, and crop performance under different climate scenarios [9, 10].
- **Enhancing Food Safety and Quality:** AI has the potential to improve food safety by detecting contaminants and ensuring quality control throughout the food production process. By letting consumers know where and how their food is produced, AI can increase food product traceability, boosting consumer confidence and ensuring food quality [11].
- **Innovation and Technological Advancements:** AI has the potential to speed research and development efforts to create novel crop kinds with enhanced resistance to pests, diseases, and climate change [12]. Automation: Robotics, AI, and other automated technologies can revolutionize agriculture and make it more labor-efficient and efficient.
- **Economic Implications:** By streamlining procedures and cutting waste, AI could lower the cost of producing food [13]. Market Competitiveness: Agricultural businesses can get a competitive advantage in the market by implementing AI, which increases productivity and sustainability.
- **Policy and Regulation:** The Informed Policy Making: AI would be most beneficial for lawmakers in the development of new rules and legislation that are not only clearer but also more efficient in helping ethical, sustainable food production. AI can also help uphold environmental and ethical standards because it can effectively check that agricultural practices comply with legal ones [14].
- **Food Safety:** Through the employment of AI driven methods, agricultural production can be improved, therefore contributing to global food security by reducing hunger and malnutrition.

- **Customer Awareness:** The application of AI in food production could raise awareness among consumers about ethical and sustainable food practices by influencing their purchasing decisions and, thus, developing a culture of sustainability.
- **Prospects for the Future:** To this end, research on the application of AI in food production is vital for ensuring that the global food system is able to meet the sustainability and resilience posed by impending challenges.

3.2 RELATED WORK

Today's agriculture is focused on sustainable practices, trying to bring in less environmental impact while maintaining productivity [6]. Among other facets of sustainability, AI improves biodiversity, reduces greenhouse effects, and maximizes resource use. It is estimated that the AI-powered precision agriculture is expected to use less water, fertilizers, and pesticides through just-in-time application where and when it is required. Along with saving resources, that really narrow approach will prevent pollutants and runoff from harming ecosystems. Other AI-enabled systems that monitor various parameters allow farmers to better control pests and diseases, thereby making typical chemical treatments less frequently applied and thus promoting greener pest control methods [7]. As such, another valuable element in sustainability is AI's ability to comprehend and predict climate effects on agriculture. This resolution can only be provided by analyzing climate data, whereby AI can offer assistance to farmers in dealing with changing conditions and making agricultural practices climate-resilient. In turn, this flexibility will be very critical for food security in an ever-hotting world. A handful of case studies have shown the successful integration of AI into food production. For instance, AI-enabled greenhouses in the Netherlands, thanks to their sensors and data analytics, optimize growing conditions for various crops, and resource use is lowered while yield increases [15]. Through weather forecasts, crop recommendations, and market prices via mobile, AI applications help smallholder farmers in India improve decision-making and stabilize income. For instance, they can use computer vision and machine learning to monitor dairy cows. By analyzing the health and behavior of cows, productivity and welfare are enhanced. Not only do they improve the welfare of animals, but they also increase the productivity of farms, as technology is able to detect early signs of illness and malnutrition. AI has enormous potential for the food production industry, although there are still obstacles to be overcome to enjoy these benefits fully: social and technological barriers, such as farmers and consumers' reluctance to change, necessitate the requirement of optimized and trusted data. Furthermore, for large-scale implementations, it is essential to ensure compatibility of different AI platforms and systems.

Standardizing data formats and protocols from the field to the fork would allow for more straightforward integration of AI tools into food production processes. Moreover, for AI solutions to be useful, organized, and widely acceptable, it will be important to collaborate with technology developers, farmers, and legislators, among others. As technology develops and potential applications appear, it is likely that AI in food production will also be enhanced [8]. Future models for AI would incorporate enhancements in food delivery and processing and may predict and respond to complex biological and environmental interactions. By taking into account the ethical and sustainability concerns, AI could also play an important role in transforming food systems to face 21st-century challenges. This would involve closely examining various questions regarding the ethical and sustainable production of food: impact on the environment, food safety, supply chain optimization, and resource management [16]. AI can be used to address the same issues: Random Forests and Gradient Boosting Machines: Used for predicting crop yield, climate conditions, and pest outbreaks, these models can lead farmers to make informed decisions about sowing and harvesting [17]. Genetic Algorithms: This model can help maximize crop yield, minimizing environmental damage by utilizing water and fertilizers efficiently [18]. Linear Programming: This would aid transport emissions and costs by optimizing supply chain logistics using LP, deciding on cost-effective routes and schedules. Deep Reinforcement Learning [19]: A subset of AI aiding planting, weeding, and harvesting of automated farming machinery, such as robots and drones. The agents select the best course of action that

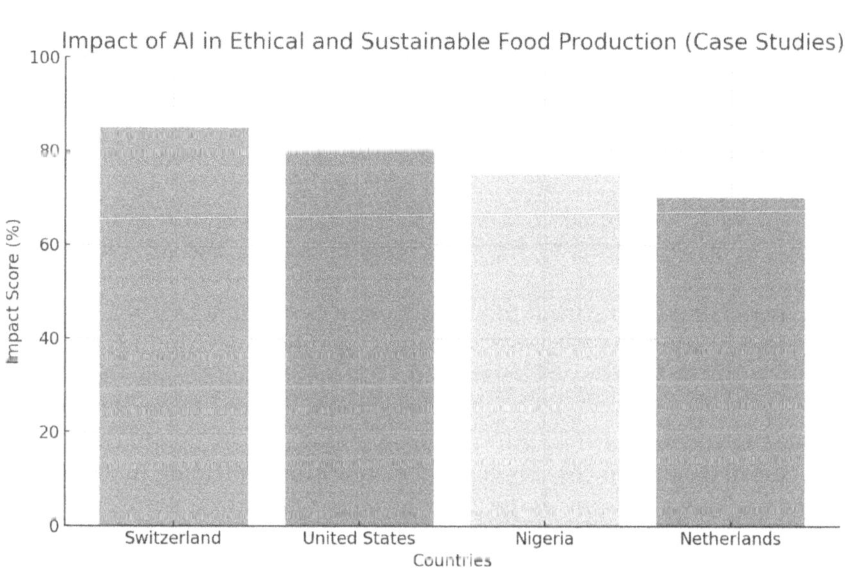

Figure 3.3 Case studies across different countries.

maximizes sustainability and efficiency [20]. Multi-Agent Systems: They can be used to simulate how different stakeholders—farmers, suppliers, and consumers—jointly strategize on achieving social, environmental, and economic objectives. Convolution Neural Networks (CNNs): CNNs are wonderful in conferring images for use in maturity measurements of crops, insect detections, and plant health monitoring [21]. Object identification algorithms (like YOLO, Faster R-CNN): These algorithms support the automated quality control of foods for precision agriculture since they are able to recognize and place individual items into images.

3.3 CASE STUDIES ON THE ROLE OF AI IN ETHICAL AND SUSTAINABLE FOOD PRODUCTION

The global agriculture is on the verge of transformation through Ai with ethical and sustainable modes of production. Many of these countries have successfully used AI-based technologies to address problems in agriculture, reduce environmental footprint, and improve food security. Such case studies have an excellent opportunity to add insight into the potential of AI in sustainable food systems. Precision agriculture driven by AI in Switzerland has also given great impetus to sustainability. Farmers would depend on algorithms from the data collected about soil, weather patterns, and health of the crops to give inputs on more efficient availability of resources. Such data-driven information would reduce water consumption, pesticide use, and increase productivity, as well as enable sustainable, environmentally-friendly farming practices. The provision of AI-embedded agricultural methods has been advocated by the government of Switzerland, as it has shown a greater potential achievement of objectives that lead to sustainable food consumption. AI has been used in the US by many companies to improve food quality management and decrease food waste. Sophisticated AI systems scanning facilities would monitor hygiene and safety in production, while their quality assurance would be automated. For example, AI can be programmed to recognize contamination and automate quality control checks, reducing human error and enhancing efficiency. That way, it contributes to ethical food production by making sure that safe and high-quality products get into consumers' hands while minimizing waste of resources. In Nigeria, AI applications are being used for the espousal of sustainable agricultural practices on smallholder farms. Using AI, farmers are able to access real-time information about the weather, soil conditions, and pest management. By relying less on traditional farming practices and better managing resource usage, these farmers can become more productive and reduce the negative influences of farming on the environment. This case study thus illustrates how AI is being used to assist small-scale farmers in developing countries to resolve food security issues.

One important study from the Netherlands considered the ethics of AI in agriculture. The researchers considered the use of AI robots in the agri-food sector and outlined the need for these technologies to be developed in an ethical manner. The study enumerated numerous concerns, including data-related privacy, displacing labor, and abuse of AI. On the contrary, it also mentioned that AI can provide a greater good toward sustainable food production through higher efficiency and less waste if applied responsibly. Figure 3.3 shows the AI impact on ethical and sustainable food production across different case studies. It discusses different areas of focus like precision agriculture, food quality control, sustainable agriculture, and ethical application of AI in Switzerland, the United States, Nigeria, and the Netherlands.

These case studies demonstrate AI's immense potential in confronting the ethical and sustainable issues presented in food production. The strategies for responsible AI deployment in agriculture can be articulated by fellow policymakers, researchers, and industry actors, drawing on the global best practices. Embedding the perspectives from these case studies in your research will yield a consolidated understanding of how AI can foster the development of a sustainable and ethical food system.

3.4 PRACTICAL IMPLEMENTATION STEPS

3.4.1 Data gathering and preprocessing

The first and foremost step toward the viable application of AI solutions for ethical and sustainable food production involves the collection of diverse datasets from multiple domains. Agricultural data—including information on crop yield, soil health, weather conditions, water use, and population of pests—can be obtained from agricultural databases, government reports, and satellite imagery [22]. Supply chain data are used to monitor the flow of food products from the farm to the table through transport, storage conditions, and retail logistics. This information can come from logistics companies, RFID tags, and IoT sensors [23]. Understanding of consumer behavior and preferences is derived from survey responses and studies, analysis of social media, and sales trends. Data on ethical and sustainability metrics, such as fair labor practices, carbon-and-water footprints, and biodiversity impacts, could come from NGOs, sustainability reports, and certification bodies [24]. Data collection methods depend on the context. The use of sensors and IoT devices is prevalent in agriculture and storage for monitoring environmental data in real time and checking food quality. Qualitative data can be collected using research and interviews with farmers, supply chain managers, and consumers [25]. Web scraping is the automated collection of data from online sources like market prices, weather forecasts, and consumer reviews [26]. Lastly, APIs from different platforms give structured datasets on agriculture, supply chains, and market trends [27].

Once collected, data preprocessing is essential to improve the quality and usability. Data cleaning consists of treating missing values through imputation or removal depending on significance, identifying outliers, and treating them using statistical techniques or machine learning algorithms and filtering any excessive or irrelevant information [28]. Numerical data are normalized to a common range (for example, 0 to 1) for compatibility, while categorical data are converted into numerical forms through techniques such as one-hot or label encoding. Then dimensionality-reducing algorithms like principal component analysis are applied to make data sets simpler by eliminating repeat features. Finally, in the step, when different data sources are integrated, datasets from various sources are merged to create a comprehensive dataset by joining tables, integrating API responses, or synchronizing data collected at different times, particularly for time-series analysis [29]. Data augmentation further enriches the dataset, especially when limited data are available. This involves generating synthetic data [30] or employing resampling techniques, such as oversampling or undersampling, to balance class distributions in classification tasks. Lastly, feature engineering is conducted to enhance analysis by deriving new features, such as crop yield per unit area or average storage temperature, and selecting the most relevant features to reduce model complexity and improve performance. By following these steps, a robust and well-prepared dataset is created, forming the foundation for successful AI applications in sustainable and ethical food production.

The graph in Figure 3.3 illustrates the hypothetical impact levels of AI across various categories in ethical and sustainable food production. Food waste reduction and climate impact are shown to have the highest potential impact (8–9), followed by categories like precision agriculture, ethical standards, and consumer awareness (7–8). Alternative proteins have the lowest relative impact (5), emphasizing a varied influence of AI in different aspects of sustainable practices.

3.4.2 Model development

Choose suitable AI models for specific situations: For image analysis applications, the corresponding applications on time series should employ long short-term memory networks (LSTMs). Models were trained and validated on historical data [11]. Since LSTMs, where the "L" stands for "long," are a type of recurrent neural network (RNN) that can learn and remember long-term dependencies, they are well-suited for time series prediction. To be able to recognize patterns through time in lengthy data sequences, they have been designed to overcome the vanishing gradient problem, which affects conventional RNNs.

Important Elements of LSTMs: Cell State: It is the memory that retains information as a long-term thing. Gates: The Forget Gate determines which cells of data will be erased from their state. The input gate selects the values of the input to be fed into the cell state. It places the output gate in the cell state. Production for the Continuously Sustaining (CS) and Ethically Validated foods: First, let us assume where you are going to observe the organic fruits and vegetables supply chain in compliance with both sustainable and moral standards. The temporal data consist of numerous variables, including humidity, temperature, pesticide usage, and transport conditions. Using LSTMs, future values and atypical behavior over these time-series data can be predicted operatively. Exemplifications: The objectives are as follows: Pesticide levels will be projected into the future so that they will stay within rational boundaries. Data: Historical concentration levels of pesticides, meteorological observations, and other relevant metrics. Using these data, build an LSTM model that predicts the level of pesticides in the future. Should thresholds approach potential violation, changes in practices can be made proactively to ensure compliance. Food production through CNNs: The main intended use of CNNs is image and spatial data processing. With convolving, pooling, and fully connected layers, they are especially adapted to the patterns and features recognition of images.

Principal Elements of CNNs: Convolutional Layers: Produce feature maps by filters applied to the input data. Pooling Layers: Consolidate and protect important characteristics, yet reduce the dimensionality of feature maps. Fully Connected Layers: Make use of high-level features extracted by convolutional and pooling layers in classification or regression tasks. Application on Ethical and Sustainable Food Production: CNNs can be trained by examining images and videos from farms, food processing sites, and retail stores in order to guarantee ethical and sustainable practices.

In guaranteeing ethical and sustainable food production, AI is showing great promise presently. Advanced monitoring and improvement schemes are made possible with AI tools; for instance, animal welfare monitoring on livestock farms is efficiently carried out by the use of CNNs. These AI models are fed with live-streaming videos from different parts of a farm to discern particular signs of distress, discomfort, or poor living conditions in animals. With real-time analysis, farm operators will be able to immediately verify and act upon any possible ethical concerns in the treatment of livestock. Stress-signifying behaviors—such as abnormal movement patterns, overcrowding, or signs of illness—can be recognized by the CNN, with alerts being generated for managers to follow up and take corrective action. Such technologies aid farmers in improving animal welfare as well as enhancing productivity. However, morality alone does not guarantee sustainability. For this reason, the conjunction of CNNs with LSTMs presents a more thorough and effective solution.

LSTMs are compelled to analyze time-series data, while CNNs are designed to analyze visual data from the camera and present current ethical dilemmas. Farms usually gather environmental data, like temperature,

Table 3.1 An overview of how AI contributes to ethical and sustainable food production

Aspect	*AI contribution*	*Challenges*	*Examples*
Precision agriculture	AI enables optimized use of resources like water, fertilizers, and pesticides by analyzing soil and crop data.	High cost of technology implementation for small-scale farmers.	AI-based systems like John Deere's See & Spray, which detect and target weeds selectively.
Food supply chain	Enhances traceability, reduces food waste, and predicts demand to ensure minimal overproduction.	Data availability and interoperability issues across global supply chains.	IBM Food Trust uses blockchain with AI for transparent food tracking from farm to fork.
Livestock management	AI monitors animal health and optimizes feeding schedules for ethical and efficient production.	Ethical concerns regarding excessive surveillance and invasion of privacy in farming operations.	Cainthus uses AI to monitor dairy cattle health and predict issues like lameness.
Alternative proteins	AI accelerates the research and development of plant-based and lab-grown meat alternatives.	High initial investment and consumer acceptance barriers.	Beyond Meat and Impossible Foods use AI for ingredient optimization and flavor profiling.
Food waste reduction	AI predicts spoilage and improves inventory management to minimize waste at production and retail levels.	Limited adoption in underdeveloped regions due to cost and infrastructure gaps.	Too Good To Go app uses AI to connect surplus food to consumers at reduced prices.
Climate impact	AI predicts climate change impacts on agriculture and suggests adaptive measures to sustain production.	Model accuracy depends on the quality of climate and regional agricultural data.	ClimateAI helps farmers plan for climate-resilient crops and strategies.
Ethical standards	Ensures compliance with fair labor practices, animal welfare, and sustainability certifications.	Lack of globally standardized ethical guidelines for AI implementation.	Fair Trade AI tools monitor and certify ethical farming practices.
Smart irrigation	AI optimizes water usage through predictive analytics, reducing wastage and enhancing efficiency.	Dependence on reliable IoT networks and sensors.	Netafim's smart irrigation systems use AI to optimize water delivery based on weather data.
Consumer awareness	AI-powered apps educate consumers on sustainable food choices and ethical practices.	Balancing transparency with oversharing proprietary data.	Apps like Yuka provide sustainability scores for products based on AI-analyzed ingredients.

humidity, feed quality, water consumption, and pesticide levels, by means of sensors or IoT devices. These data are run through the LSTMs to analyze trends and predict forthcoming conditions that would affect sustainability. Take, for example, the introduction of risks to the health of animals due to high-temperature patterns and low feed quality trends. Such risks can lead to LSTM-advisory responses to either adjust feeding schedules or modify certain ventilation systems. The real magic starts when you combine CNNs and LSTM networks; such an AI system offers real-time monitoring with answers to environmental inquiries. Here, for example, the CNN could identify immediate issues such as overcrowding in a pen, while on the other hand, the LSTM is predicting the effect of high humidity in that pen on the animals over the next week. Thus, these instruments will enable farm managers to address ethical issues arising both in the short run and in a sustainable manner in the long run. Such integration promises that farms not only implement the ethical treatment of animals but they also curb resource utilization and environmental factors for the establishment of sustainability in practice. This kind of integrated approach gives a composite view of farm operations so that they do not deviate from either ethical treatment of livestock or environmental sustainability. Such a system uses CNN outputs for immediate issues like animal distress detection and LSTM predictions for future challenges in balancing effective productivity with ethical and sustainable ways of farming. This robust AI system, in all, converges toward the wider goals of ethical and sustainable food production, not just for the benefit of livestock but also for the environment and consumer expectations from responsible farming.

AI-Driven Production Practices: While a comprehensive integration is needed, a governing role for such a system would require generalized activities for the ethical treatment of livestock and environmental considerations. For some support in immediate issues such as distress detection in animals, CNN-based predictions are used, while some predictions offered by the LSTM about issues that lie in the years to come would balance production efficiency with the pathways to ethical and sustainable agriculture. All of these AI systems ensure a greater good and not simply for the well-being of the livestock, but with real benefits for the environment and in accordance with consumer expectations.

In fact, for the success of AI, the deployment and integration of AI to existing farming operations and supply chains are of ultimate importance. Development of AI models and their integration into existing frameworks would involve interoperable and easily understandable dashboards and interfaces to simplify the understanding of AI-generated information for all stakeholders. Continuous monitoring and evaluation of these AI models are mandatory for the evaluation of the cumulative performance and its benefits to the ethics and sustainability practices. Stakeholders' feedback should be collected and, based on that, areas for improvement identified for enhancing the models to render them more efficacious with time. Morality has to be the axiom every time development and integration of any AI systems are considered-breaching standards for taking any government data and placing it on the web openly. Setting aside the forced-to-purchase data standards, the AI is both a driver of the awareness on related news and a plug for adaptation based on returning data. Stakeholder involvement is especially important for addressing the most critical concerns from stakeholders and to engage their ideas in redesigning the system to create that trust and collaboration while maintaining good ethical AI practices.

Some of the transformative AI based frameworks in the realm of ethical and sustainable food production (AI) across various sectors is listed in Table 3.1. AI based precision agriculture can optimize the use of water, fertilizers, and pesticides on a case-by-case basis. However, its high adoption cost makes it more accessible to large agribusinesses than to smallholder farmers. AI brings considerable potential to the food supply chain by enabling traceability, waste reduction, and demand prediction. Data sharing is a challenge between global systems, as exemplified by the blockchain-based tracking of IBM Food Trust. AI models monitor animal health and weight management, coupled with schedules, but with hardly any ethical dilemmas, like in the case of Cantus' dairy cattle health monitoring. Alternative proteins are possible by accelerating AI-developed foods through plant proteins or lab-grown foods; however, high costs and consumer acceptance are dominant in such a venture, for example, ingredient optimization is being employed by movers like Beyond Meat. AI (should be noted) also reduces food loss; predicting spoilage for good inventory management is again in full throttle. Yet, limited to no adoption exists qualitatively in less developed regions, as with the likes of the Too Good to Go app. It stands as the backbone for climate-resilient agriculture, predicting climate risks in tandem with adaptive strategies, as seen by Climate.

3.5 CONCLUSION

AI has the potential to transform food systems, working a fine single line that takes moral and environmental objectives along with the bounds of its potential achievements. AI has the capacity to actually utilize data analytics, automation, and prediction modeling in many food production aspects, also with low technology, ranging from waste minimization, supply chain visibility terminology, and resource optimization. For instance, considering the ecological costs, farmers and producers might apply data-based decision-making to prevent environmental damage, as pesticide, fertilizer, and water utilization stand for the effective operation of agriculture. The use of AI can promote eco-growth, while it relies on increased pinpoint farming, diagnosing and treatment, and monitoring plants and livestock accordingly. Such an operational scheme for forest utilization reverses the degradation of farm chemicals and conservation of natural resources. On the other hand, AI plays a crucial ethical role in how food is produced. Enhanced traceability and transparency mechanisms along the supply chain ensure that food is humanely produced along with ethical labor rights and fair trade under the aegis of AI. Its monitoring setup seeks out unethical inclusions and alerts just in time to ensure timely interventions and the bringing of the food closer to ethicality. In the space of food waste, AI lessens it by managing its inventories optimally, predicting all demand with precision, while hitting the mark and integrating logistics to ensure that food reaches customers just in time. Controlling food waste is the most important element for a sustainable food system, as it saves resources and the environment from destruction in food production. But every possible care should be taken when incorporating AI in food production, certainly in the context of potential ethical and societal implications. Addressing concerns pertinent to ethical considerations, such as data privacy, the digital divide, and concerns of loss of jobs by introducing robust measures that channel AI benefits to society would be desirable. Good governance and leadership are the two main pillars that are strongly required in observing the development and application of AI in that very manner in which it agrees with social values and objectives. AI, in continuation, should also be managed prudently, pretty much in conflict with ethical considerations and scientific advancements. AI presents good prospects for use toward all sustainable and ethical recruitment. Wouldn't that be something? AI would become a magnetic core of effort upon which those nations would survive to lead the way out of the global arena of wickedness for a fair and sustainable food system.

REFERENCES

1. P. Akhtar, J. G. Frynas, K. Mellahi, and S. Ullah, "Big data-savvy teams' skills, big data-driven actions and sustainable performance," *Prod. Plan. Control*, vol. 30, no. 7, pp. 567–581, 2019.
2. S. A. Bhat, N. Huang, A. Gulati, A. Kumar, and V. Athreya, "Application of artificial intelligence and Internet of Things in agriculture," *Agronomy*, vol. 11, no. 6, p. 1085, 2021.
3. R. Bogue, "The role of artificial intelligence in sustainable agriculture," *Ind. Robot Int. J.*, vol. 41, no. 4, pp. 391–396, 2014.
4. S. Bonny, "Corporate social responsibility (CSR) in the agribusiness: CSR activities in the sustainable food production," *Agron. Sustain.* Dev., vol. 37, no. 5, pp. 1–8, 2017.
5. K. Bronson, "Smart farming: Including rights holders for responsible agricultural innovation," *Technol. Innov. Manag. Rev.*, vol. 8, no. 2, pp. 7–14, 2018.
6. K. Bronson and I. Knezevic, "Big Data in food and agriculture," *Big Data Soc.*, vol. 3, no. 1, pp. 1–5, 2016.
7. A. Buschulte, M. Kayser, and P. Weichselbaum, "Ethical perspectives on artificial intelligence (AI) in precision farming," *Precis. Agric.*, vol. 21, no. 6, pp. 1197–1209, 2020.
8. J. P. Chavas and C. Nauges, "Uncertainty, learning, and technology adoption in agriculture," *Appl. Econ. Perspect. Policy*, vol. 42, no. 1, pp. 42–53, 2020.

9. C. Costa, F. Antonucci, F. Pallottino, J. Aguzzi, Da-Wen Sun, and P. Menesatti, "Shape analysis of agricultural products: A review of recent research advances and potential application to computer vision," *Food Bioproc. Technol.*, vol. 6, no. 3, pp. 566–572, 2013.
10. S. Wolfert, L. Ge, C. Verdouw, and M. J. Bogaardt, "Big data in smart farming–a review," *Agric. Syst.*, vol. 153, pp. 69–80, 2017.
11. M. Das and N. K. Arora, "Artificial intelligence: A boon for sustainable agriculture," *Sustainability*, vol. 12, no. 14, p. 5248, 2020.
12. M. De Clercq, A. Vats, and A. Biel, "Agriculture 4.0: The future of farming technology," *World Gov. Summit Rep.*, 2018.
13. U. Deichmann, A. Goyal, D. Mishra, and T. Vishwanath, "Will digital technologies transform agriculture in developing countries?" *Agric. Econ.*, vol. 47, no. S1, pp. 21–33, 2016.
14. A. Elghannam, M. Escribano, F. J. Mesias, and M. Escribano, "Food traceability systems: How knowledge and communication issues challenge user confidence in pork along the supply chain," *Food Control*, vol. 71, pp. 159–166, 2017.
15. R. Finger, S. M. Swinton, N. El Benni, and A. Walter, "Precision farming at the nexus of agricultural production and the environment," *Annu. Rev. Resour. Econ.*, vol. 11, pp. 313–335, 2019.
16. A. V. Goodchild and B. P. Mullan, "Artificial intelligence in agriculture," in *Artificial Intelligence in Agriculture*, Cham: Springer, 2020, pp. 1–23.
17. A. Grill, "The role of artificial intelligence in food production and sustainability," *J. Clean.* Prod., vol. 264, p. 121628, 2020.
18. A. Kamilaris, A. Kartakoullis, and F. X. Prenafeta-Boldú, "A review on the practice of big data analysis in agriculture," *Comput. Electron. Agric.*, vol. 143, pp. 23–37, 2017.
19. K. E. Giller et al., "The future of farming: Who will produce our food?," *Food Secur.*, vol. 13, pp. 1073–1099, 2021.
20. L. Klerkx, E. Jakku, and P. Labarthe, "A review of social science on digital agriculture, smart farming and agriculture 4.0: New contributions and a future research agenda," *NJAS-Wageningen J. Life Sci.*, vol. 90, p. 100315, 2019.
21. K. G. Liakos, P. Busato, D. Moshou, S. Pearson, and D. Bochtis, "Machine learning in agriculture: A review," *Sensors*, vol. 18, no. 8, p. 2674, 2018.
22. H. S. Moghadam and K. Kalantari, "The role of artificial intelligence in sustainable food systems: A review," *Sustainability*, vol. 13, no. 14, p. 7669, 2021.
23. Z. Niu, Z. Zhang, and L. Zhang, "Impact of artificial intelligence on green supply chain management," *J. Clean. Prod.*, vol. 242, p. 118477, 2019.
24. M. Perez, L. Kurkalova, and S. M. Swinton, "The role of machine learning in sustainable agriculture," in *Data Science for Agriculture and Natural Resource Management*, Cham: Springer, 2017, pp. 37–57.
25. K. Poppe, S. Wolfert, C. Verdouw, and A. Renwick, "A European perspective on the economics of big data," *Farm Policy J.*, vol. 10, no. 1, pp. 11–19, 2013.
26. S. Rotz and E. D. Fraser, "Resilience and the industrial food system: Analyzing the impacts of agricultural industrialization on food system vulnerability," *J. Environ. Stud. Sci.*, vol. 5, no. 3, pp. 459–473, 2015.
27. M. Ryan and E. Shook, "The role of AI ethics in sustainable development," *Nat. Mach. Intell.*, vol. 1, no. 12, pp. 566–570, 2019.

28. A. Schrijver and W. van Hoof, "The role of big data in sustainable food production and distribution," *Food Bioproc. Technol.*, vol. 9, no. 6, pp. 1032–1041, 2016.
29. G. D. Stone, "The anthropology of algorithms and algorithms of anthropology," *Am. Anthropol.*, vol. 122, no. 1, pp. 22–35, 2020.
30. G. Tripathi and R. Lal, "Intelligent solutions for smart farming: A comprehensive review," *IEEE Access*, vol. 6, pp. 80188–80207, 2018.

Chapter 4

Innovations, applications, and future prospects in the agriculture sector through emerging technologies

Smriti Singh, Salman Khursheed Ahmad, and Km Ikra

4.1 INTRODUCTION: TRENDS IN AGRICULTURE

Figure 4.1 shows the use cases of smart agriculture. Smart spraying, disease prediction and detection in plants, smart irrigation, and smart soil fertilization are some use cases of smart agriculture.

4.1.1 Precision agriculture

It's the science of enhancing the yields of crops so as to support them through the management of decisions by sensors with high-tech tools of analytics. As an emerging technique, precision agriculture employs inputs that are optimized in nature such as fertilizers and water, so as to increase the quality, productivity, and yield in the fields of agriculture. It depends on a significant amount of the information as per the conditions of the crop means the health during the season of growth, collected at the spatial resolution which is high. Irrespective of the source of data, the primary goal of precision agriculture is to assist the farmers in managing the operations [1]. Precision agriculture has proven to utilize the data from many sources in order to improve yields and enhance the effectiveness of the various management strategies, like application of fertilizer, management of irrigation, and the usage of pesticides. While the definitions of precision agriculture may change, the process generally refers to the tools that are applicable to the scale of a local subfield. Remote sensing is being recognized as an important and crucial source of information for very precise agriculture; however, it is being adopted slowly for many reasons [1, 2]. However, emerging technology enhancement, mainly the presence of high-resolution satellite imagery and the rapid development of unmanned aerial vehicle technology, indicates that the use of remote-sensing data in precision agriculture is very likely to increase significantly in the future. Figure 4.2 represents the precision farming cycle [3].

 DOI: 10.1201/9781003494782-4

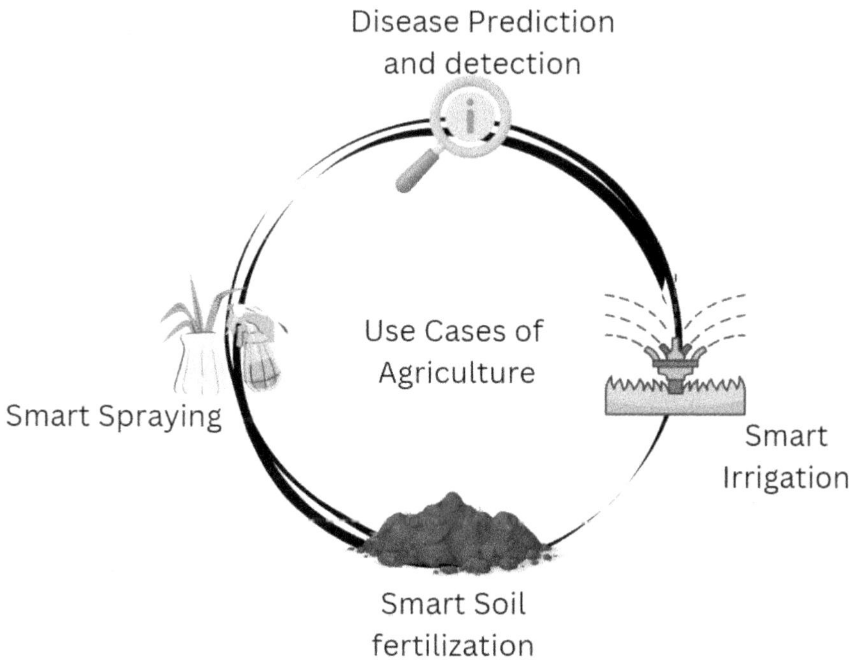

Figure 4.1 Use case of agriculture.

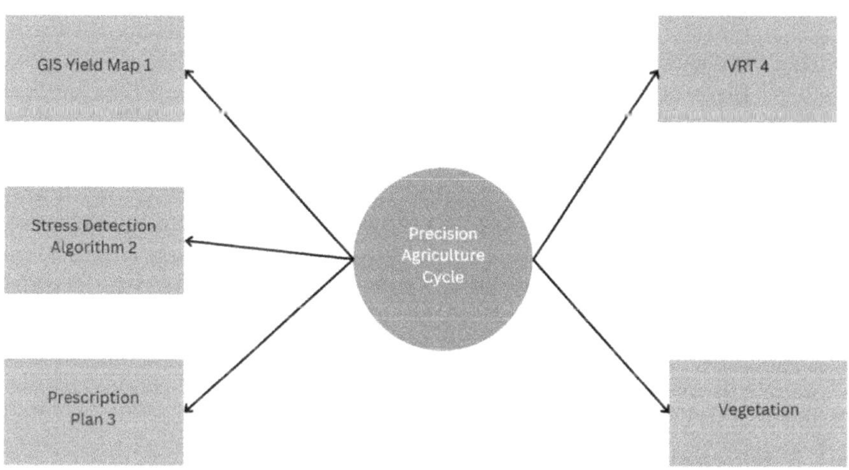

Figure 4.2 Precision agriculture cycle.

4.1.2 Sensors

A sensor is an electronic device that can detect a specific kind of information from the environment and transform it into an electrical signal or another form that is required for the information to get the output. Sensors are majorly programmed to meet the needs of transmission of information, storage, processing, display, recording, and control. In the field of agriculture, sensors are mainly used in order to collect data about the environment or the plants, which can further be converted into electrical signals that can be used for analysis purposes. With the Internet of Things (IoT) technology, the sensors are increasing day by day in terms of manufacturing and can be applied across various industries in the field of agriculture. The IoT technology provides the sensors with the ability to gather information from multiple locations and then transmit it to the central hub, where it can be further utilized for making agricultural decisions that are informed [3, 4].

On the basis of the type of sensor, it can be placed in the soil where the probes can read the data and then convert it into electrical signals, or we can say readings, so that it can be interpreted by humans.

A. Best Types of Sensors

Here are several types of sensors in agriculture are beneficial for a wide range of crops, fields, and situations. Given below are six of the best types of sensors

- **Optical Sensors:** This kind of sensors uses the light in order to measure the data. These sensors can be mounted on the vehicles, drones, or robots which are capable of mobility to collect the data. Optical sensors cause the soil to reflect light, providing the farmers with insights of soil and the plant color.
- **Electrochemical Sensors:** This kind of sensors collects the chemical data related with the soil by detection of specific ions. It supplies information about the pH levels and the content of soil nutrient. The main goal of this kind of the sensors is to obtain the highly specific information of the soil in order to facilitate the informed decision-making.
- **Mechanical Sensors:** These kinds of sensors are used to measure the compression in soil or the mechanical resistance. They document the forces being encountered during processes like crushing, cutting, and displacing the soil [5].

4.1.3 Blockchain

This technology plays a major role in field of agriculture; it enhances food safety by tracing the information throughout the food supply chain.

There are various applications of blockchain in agriculture, listed as below:

- Intelligent Farming
- Food Supply Chain
- Agricultural Insurance
- Agricultural Product Transactions
- Smart Agriculture
  All of the procedures in the old method of controlling smart technology are often centralized, which can lead to different mistakes and distortions in the data collected. Figure 4.3 is the depiction of blockchain in agriculture.
- **Smart Farming Model:** The value and potential of the blockchain technology in agriculture result in the development of the several

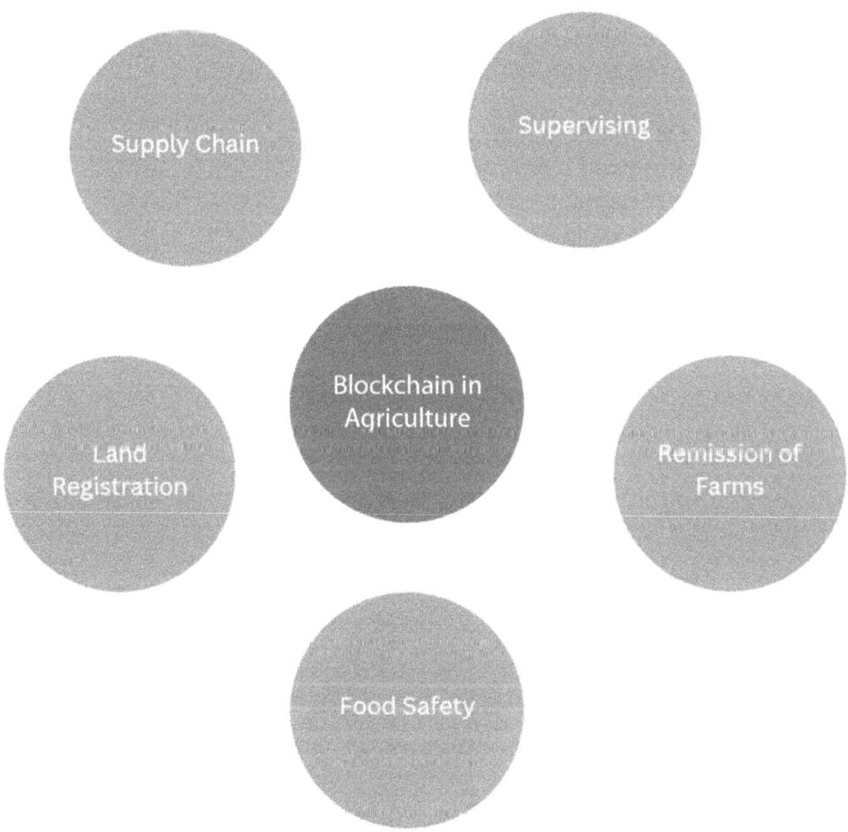

Figure 4.3 Blockchain in agriculture.

models of smart farming, which combine the pros of this technology with those of other IoT sensors [2].
- **Smart Farming Technology:** Organizations have started to innovate the smart agricultural technology. For example, a company has developed products that integrate smart farming technology, connecting the several multiple networks to the physical items. They created a coin-sized device to help users conduct secure transactions on a blockchain [3].
- **Food Supply Chain:** The chain of food supply has become longer and very complex than ever due to globalization. These issues place a burden on the economy and society of the customer's health.

4.1.4 Biotechnology

Biotechnology has a variety of tools and elements from the conventional techniques of breeding, bioinformatics, microbiology, molecular genetics, biochemistry, plant physiology, and molecular biology. The key biotechnology tools applicable to agricultural biotechnology are [6]:

- Conventional plant breeding
- Tissue culture and micropropagation
- Molecular breeding (or marker-assisted selection)
- Genetic engineering and genetically modified crops
- Molecular diagnostic tools.

Here, Figure 4.4 shows the use of biotechnology in agriculture.

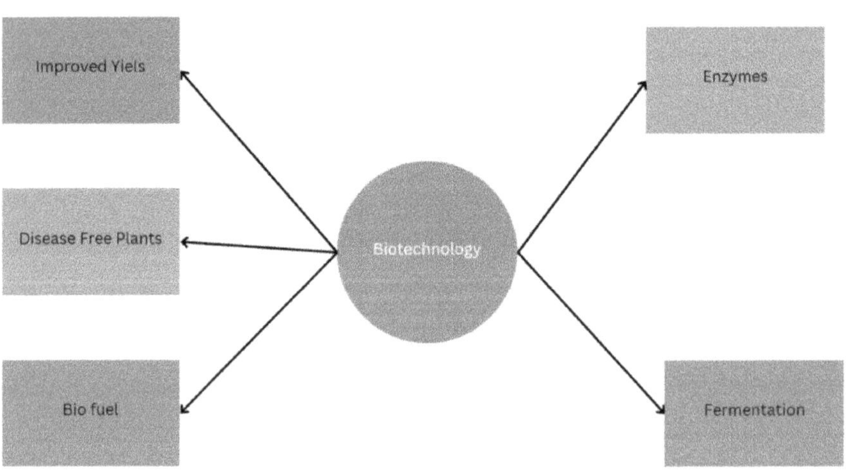

Figure 4.4 Biotechnology in agriculture.

4.1.5 Robots

For a huge industrial farm means on a larger scale, robots and automation are seen to be crucial where there is extensive land to monitor and where large quantities of crops need to be processed [7, 8].

At present, the robots are used for a number of different tasks, including:

- **Maintenance of Crops:** Robots are used as they can be programmed to perform the tasks, for instance, as weeding, pruning plants, applying insecticides or nutrients, and providing them with the irrigation requirements.
- **Seeding:** Driverless tractors and robots are gradually replacing the machinery as a traditional method for sowing the seeds.
- **Harvesting:** Due to the shortage of labor, robots are primarily utilized for harvesting ripened fruits and vegetables.
- **Palletizing:** This task, typically performed by forklift trucks operated by drivers, is increasingly automated using palletizing robots or robotic arms.
- **Packing:** Robotic arms are used to sort and pack produce by type, and they can be programmed to make the difference between the good and the bad production, ensuring only the best reaches the supermarkets.
- **Productivity:** Robots can work faster than humans and for longer periods without a decrease in productivity, injury risk, or need for breaks.
- **Workforce Guaranteed:** For the tasks that are seasonal, like fruit-picking, it can be further challenging to find the staff in time.
- **Reduction of Waste:** Automation prevents the crops from being wasted due to staff shortages or delays in transport for distribution.
- **Precision:** Robots are not prone to errors like humans, ensuring that both simple and complex tasks are performed accurately and consistently throughout the process.
- **Efficiency Cost:** The initial investment may be high, but robots operate 24/7, leading to efficient farming.

Figure 4.5 represents various applications of smart agriculture.

4.1.6 Digital farming

This method equips the farmers with timely and valuable insights and enables them to adopt the best practices and to manage their operations more efficiently, thus minimizing losses and maximizing profits. There are various technologies to provide solutions for advanced farming [1, 9].

Benefits of agricultural solutions that are digital for the processing of the food industry include the following:

- End-to-End solutions from farm to consumer
- Increased the yields due to the optimized and continuously monitored inputs

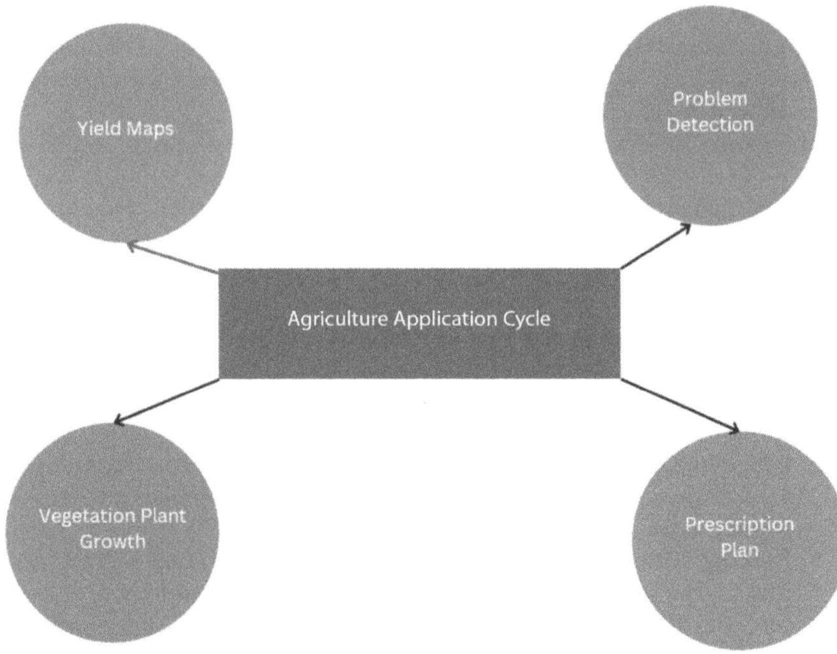

Figure 4.5 Agriculture application cycle.

- Improved the quality through compliance with the food standards and the nutrition track
- Reduced the waste via practices that are customized and can ensure a precise application of resources, by lowering production costs

4.1.7 Drones

Drones have already undergone significant transformation in the agricultural field, and they will continue to grow in the future. Figure 4.6 shows the monitoring of crops using drones.

Drones are used to monitor the health of the soil and the conditions of the field. They can produce field maps, which are accurate in nature and can include the information, helping growers detect elevation irregularities in their respective fields [8, 10].

- **Monitoring of the Plant Health:** This capability allows the farmers to monitor the crops as they grow, enabling them to a quick action in order to address any issues that may arise.
- **Planting and Seeding:** A less widespread use of drones in the field of agriculture is in planting seeds. Automation of the drone seeders

Figure 4.6 Drones in agriculture.

is mainly used in the field of forestry industries at present, but the potential for more widespread use is on the horizon.

- **Application of Spray:** Drone use to apply spray treatments is already widespread in the Southeast Asia, with South Korea using the drones for approximately 30% of their agricultural spraying.
- **Security:** Drone security is extremely useful in the area of farm management. Drones can be used to monitor the far reaches of a farm without having to waste valuable time and allow for more frequent monitoring, in areas that are hard to reach.
- **AI in Drones:** Another drone technology involves machine learning. Improving artificial intelligence (AI) in drones is very important to make them useful for smaller farmers in developing nations.
- **Pollination through Drones:** The newest uses for drones in agriculture are still in testing and development.
- **Irrigation through Drones:** There are great opportunities for drone use in agriculture. As climate change increasingly affects drought conditions, creating more efficient irrigation solutions is vital.

4.2 AI IN AGRICULTURE

AI provides local farmers with access to information and the knowledge of the tools and enables them to make more informed decisions about their practices of farming. Figure 4.7 shows smart agriculture use cases [1].

Traditional farming involves various manual processes. Implementing the models of AI can have many advantages in this respect.

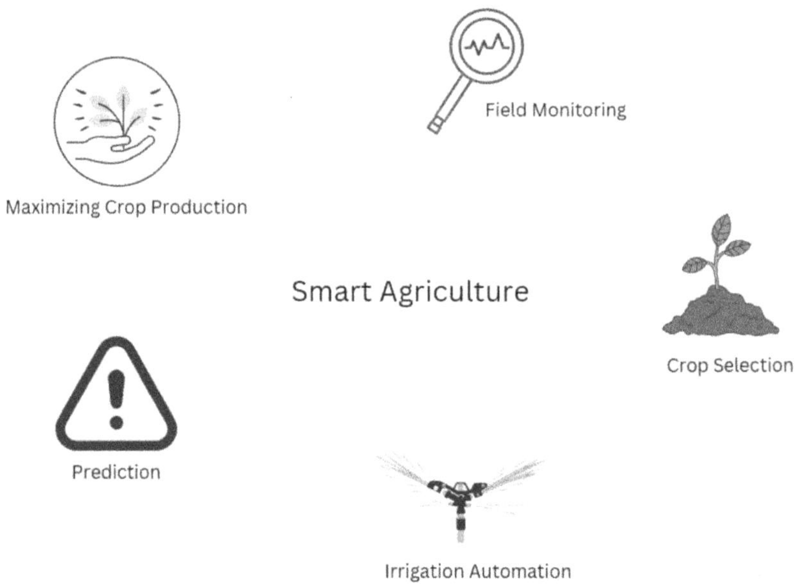

Figure 4.7 Smart agriculture.

4.2.1 Optimizing automated irrigation systems

The algorithms of AI enable the autonomous management of crops. When combined with the sensors of the IoT, which can monitor the levels of soil moisture and the conditions of weather, the algorithms can decide in real time how much water is needed for the crops [2].

4.2.2 Damage to the system of irrigation systems or leakage detection

AI uses weather and crop data to identify the areas where water is being excessively used; the algorithms can identify the patterns and the anomalies so as to indicate the leaks [3].

4.2.3 Pests and disease detection

The field of computer vision helps in detecting the presence of pests or diseases.

4.2.4 Monitoring of crop and soil

AI allows farmers to easily make the necessary adjustments by seeing the combination of nutrients in soil, which can affect the health of crops. Identification of these kinds of nutrients helps in creating a good yield of the crop [10].

4.2.5 Produce harvested sorting

The technology of computer vision helps in detecting pests as well as diseases in the crops being harvested. It can be produced based on its shape, size, and color.

Figure 4.8 represents the use of AI in agriculture.

- Promotion of economical and sustainable farming by enhancing access to finance for farmers who are smallholders
- Satellite data fuels advances in agritech
- The growing impact of cloud computing
- High-speed 5G networks connecting the global agriculture ecosystem

4.3 TRANSFORMATION OF AGRICULTURE THROUGH TECHNOLOGY

- Technology is revolutionizing the agriculture industry, leading to more efficient and environmentally friendly farming practices.
- Modern farming techniques and sustainable innovations are reshaping the way we cultivate crops and raise livestock.

4.4 THE IMPACT OF TECHNOLOGY ON FARMING

Technology has had a profound impact on the agriculture industry, revolutionizing farming practices and paving the way for more efficient and sustainable methods of crop cultivation and livestock management [3].

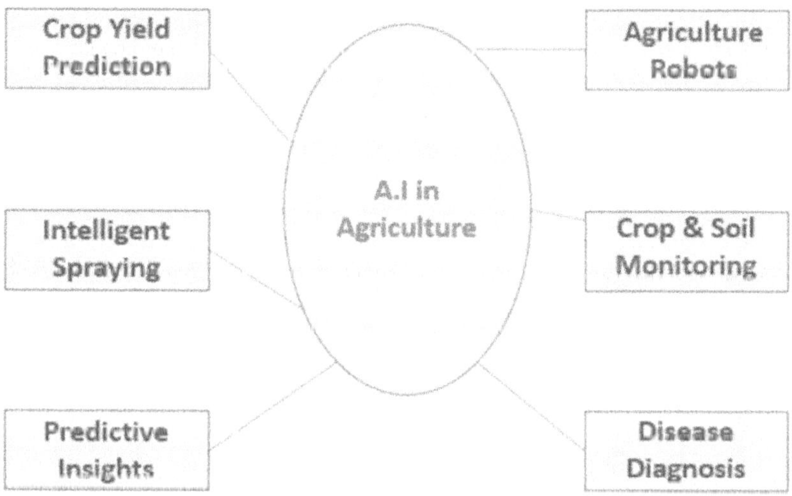

Figure 4.8 AI in agriculture.

4.4.1 Sustainable farming practices

Technology has greatly contributed to the development and adoption of sustainable farming practices. With innovative solutions such as vertical farming, hydroponics, and aquaponics, farmers can cultivate crops in controlled environments, using less land, water, and nutrients compared to traditional farming methods [7].

4.4.2 The digital transformation in agriculture

Technology is driving a digital transformation in the agriculture industry, revolutionizing traditional farming practices. Farmers are leveraging precision agriculture tools and cutting-edge agri-tech solutions to make data-driven decisions, optimize operations, and achieve increased productivity and sustainability. Key benefits of the digital transformation in agriculture include [8]:

- Improved crop yield and quality
- Enhanced resource management and efficiency
- Increased profitability and market competitiveness
- Enabled data-driven decision-making
- Reduced environmental impact

The digital transformation in agriculture is shaping the future of farming, enabling a sustainable and resilient industry.

4.5 ADVANCEMENTS IN AGRICULTURAL TECHNOLOGY

As the agriculture industry continues to evolve, advancements in technology are playing a pivotal role in driving innovation and transforming farming practices [9].

- **Digital Solutions for Agriculture**
  Digital solutions are empowering farmers with data-driven insights and decision-making capabilities. AI-powered tools analyze vast amounts of agricultural data, providing valuable recommendations for optimizing crop growth, pest control, and irrigation.
- **Precision Agriculture Tools**
  Precision agriculture tools are revolutionizing farming techniques by offering precise control and monitoring of various aspects of crop production.
- **Smart Farming Innovations**
  Smart farming innovations encompass a wide range of technologies that optimize agricultural processes and improve overall farm

management. Automation systems automate tasks such as seeding, spraying, and harvesting, reducing labor costs and increasing efficiency.

- **Modernizing the Agriculture Industry**
  The effect of technology on farming is transformative in nature. It is one of the major areas driving modernization.

4.6 SUSTAINABLE AGRICULTURE AND AGRIBUSINESS INNOVATION

The intersection of these two fields holds immense potential for addressing the challenges of food production while ensuring long-term environmental and economic sustainability [10].

4.6.1 The importance of sustainable agriculture

Sustainable agriculture targets to minimize the impact of farming on environmental activities while sustaining production and ensuring the high quality of crops. It focuses on the preservation of the natural resources.

4.6.2 Agribusiness innovation: driving sustainable practices

The agri-food companies may optimize the utilization of the resources, waste being reduced, while meeting the consumer demands for the ethically and socially responsible products of agriculture.

4.6.3 Technologies of agriculture

- Vertical farming techniques enable year-round crop production in urban areas, reducing transportation costs and carbon emissions.
- Remote sensing technologies and satellite imagery provide accurate monitoring of crop health, aiding in early pest detection and efficient resource allocation.
- Smart irrigation systems optimize water usage, reducing waste and conserving this precious resource.

4.6.4 The future of sustainable agriculture and agribusiness

Using digital solutions and harnessing the power of agriculture technology, farmers can achieve a delicate balance between productivity and sustainability, while meeting the increasing demand for responsibly sourced food.

4.6.5 Precision agriculture solutions for the modern farmer

When it comes to modern farming technologies, precision agriculture solutions are at the forefront of revolutionizing the agriculture industry.

4.6.6 Advanced data analytics for optimal decision-making

"Precision agriculture solutions empower farmers to make data-driven decisions and take proactive measures to protect their crops."

4.7 THE AGRITECH REVOLUTION

Vertical farming is another innovation that is revolutionizing the way we grow crops. By stacking multiple layers of plants in indoor environments, vertical farms optimize land use and minimize water consumption [2].

4.7.1 Benefits of technology in farming

By embracing sustainable agriculture technology solutions, farmers can optimize their operations, increase productivity, and contribute to an eco-friendly and efficient agricultural sector.

4.7.2 Improved efficiency

Automated machinery and smart systems streamline various processes, reducing manual labor and saving time [2, 3].

4.7.3 Reduction of resource waste

With the utilization of the précised tools of agriculture and the data-driven insights, farmers can well optimize the use of resources such as fertilizers, water, and pesticides.

4.7.4 Improvement of crop quality

Technology plays a vital role in the field of agriculture, mainly in the improvement of the quality of crops, which directly affects the market value and consumer satisfaction. Figure 4.9 represents the crop monitoring through drones.

4.7.5 Increased profitability

Adoption of the agricultural technologies that are modern in nature can lead to improved profitability for the farmers. By doing operations of optimization, reduction in waste, and improved quality of crop, farmers can achieve higher yields and revenue.

Figure 4.9 Crop quality in agriculture.

4.7.6 Challenges in the implementation of the agricultural technology

The implementation requires a collaborative effort from various stakeholders, including government bodies, research institutions, and the private sector organizations.

4.8 CONCLUSION

The integration of technologies into the agriculture industry brings about a significant change, with innovative farming techniques, digital solutions, and precise agriculture tools; using these, farmers can optimize operations, enhance sustainability, and improve productivity with better yields of crops. By providing suitable funding, research grants, and regulations, the governments empower farmers and agribusinesses in order to embrace the practices that are sustainable in nature and leverage the latest advancements in technology related to agriculture.

REFERENCES

[1] U. Kose, V.B.S. Prasath, M.R.H. Mondal, P. Podder, S. Bharati, *Artificial Intelligence and Smart Agriculture Applications*, 2022, https://doi.org/10.1201/9781003311782.

[2] N. Wang, N. Zhang, M. Wang, Wireless sensors in agriculture and food industry – recent development and future perspective, *Comput. Electron. Agric.* 50 (2006) 1–14, https://doi.org/10.1016/j.compag.2005.09.003.

[3] H. Agrawal, R. Dhall, K.S.S. Iyer, V. Chetlapalli, An improved energy efficient system for IoT enabled precision agriculture, *J. Ambient Intell. Hum. Comput.* 11 (2020) 2337–2348, https://doi.org/10.1007/S12652-019-01359-2.

[4] X. Zhang, X. Fu, Y. Xue, X. Chang, X. Bai, A review on basic theory and technology of agricultural energy internet, *IET Renew. Power Gener.* 18 (2024) 1318–1331, https://doi.org/10.1049/RPG2.12808.
[5] S. Santiteerakul, A. Sopadang, K.Y. Tippayawong, K. Tamvimol, The role of smart technology in sustainable agriculture: a case study of wangree plant factory, *Sustainability* 12 (2020) 1–13, https://doi.org/10.3390/su12114640.
[6] National Mission for Sustainable Agriculture (NMSA), retrieved from https://nmsa.dac.gov.in (Accessed 28 February 2023).
[7] Indian Ministry of Agriculture and Farmers Welfare (2018), retrieved from https://pib.gov.in/PressReleaseIframePage.aspx?PRID=1515792 (Accessed 28 February 2023).
[8] P. Shetty, K. Manorama, M. Murugan, M. Hiremath, Innovations that shaped Indian agriculture: Then and now, *Indian Journal of Science and Technology* 7(8) (2014) 1176–1182.
[9] D.K. Pandey, P. Kumar, e-Readiness of teachers in higher agricultural education of North Eastern hill states of India, *The Indian Journal of Agricultural Sciences* 89 (7) (2019) 1212–1216.
[10] R. Chand, P. Kumar, S. Kumar. Total factor productivity and returns to public investment on agricultural research in India, *Agricultural Economics Research Review* 25 (2012) 181–194.

Chapter 5

Next-gen agriculture

Resolving issues with 5G and emerging technologies on sustainable solutions

Sharma Ji, Ayushi Prakash, Vishal Jayaswal, and Malik Shahzad Ahmed Iqbal

5.1 OVERVIEW OF AGRICULTURAL DIFFICULTIES

Various issues in agriculture affect the economy, environment, and food. Shifting weather patterns, including increased drought and changing growing seasons due to climate change, disrupt agricultural cycles and contribute to declining yields. Intensive agriculture, deforestation, and erosion have led to soil degradation, which, in turn, has resulted in nutrient depletion, making it harder to maintain high output levels [1]. According to the World Bank, agriculture requires substantial amounts of water, especially irrigation, which puts a strain on the already limited supplies, making water scarcity an even larger issue. Moreover, climate change is expanding the range of diseases and pests, and pesticide use is resulting in reduced resistance of crops. Fluctuating grain prices, skyrocketing operating costs, and volatile economies put immense economic pressure on farmers [2]. The sourcing of labor continues to be a challenge, especially in rural regions, where difficult conditions, depopulation, and low pay have driven farmers away from rural areas. Farmers without ready access to vital terminals, water, and modern storage systems are highly unlikely to realize higher profits and com petitiveness, as characterized by small-scale farmers [3]. Moreover, grabs displace many farmers, and insecurity of land tenure discourages investments in more sustainable practices. Also, inconsistent government policies and insufficient assistance, particularly to smallholders who lack access to financing or subsidies, compound the difficulty of farming. Moreover, runoff flows into water bodies while monoculture and the use of pesticides deplete ecosystems by endangering biodiversity. We need a multifaceted approach to these challenges that includes climate-proof crops, sustainable practices, better water management, finances, and farmer-friendly policies. Figure 5.1 outlines better farming practices, which is the goal of the framework. It enables processes like data collection and analysis through automatic programs. Furthermore, Internet of Things (IoT) helps provide sensors to remotely monitor the conditions of the fields as well as other agricultural aspects.

DOI: 10.1201/9781003494782-5

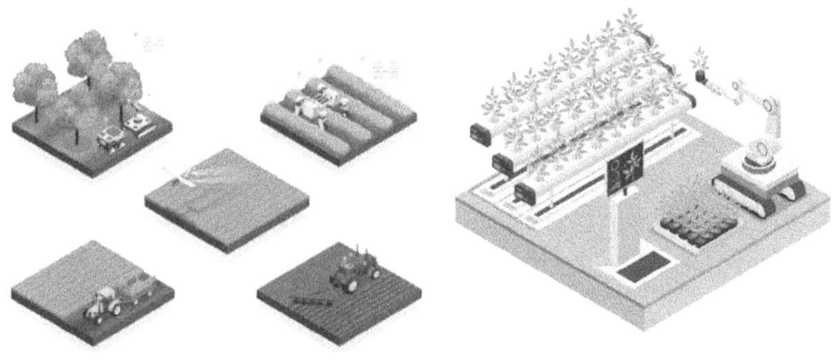

Figure 5.1 Smart farming process.

In a bid to ensure improved productivity and preservation, the said system complements marketing policies, logistics, and agricultural financing and insurance.

There are nine major sections in this chapter. A summary of traditional farming practices and the main challenges facing modern agriculture is included in Section 5.1's review of agricultural challenges. The difficulties and obstacles, such as inadequate infrastructure, financial problems, and a lack of education and training, are the main topic of Section 5.2. The role of 5G technology in agriculture is examined in Section 5.3, with an emphasis on enhanced connection, data transfer, and real-time management and monitoring. Precision farming, drones, robots, IoT applications, and advances in artificial intelligence (AI) and machine learning are just a few of the innovative agricultural technologies covered in Section 5.4. Precision farming is highlighted in Section 5.5, which also discusses its definition, importance, and 5G-supported techniques. The integration of 5G and sustainable practices is covered in Section 5.6, with particular attention to resource and water management and soil health monitoring. Section 5.7 examines case studies of 5G application where developing technology helps anticipate agricultural innovations while supplying additional competencies at international success levels. Section 5.8 analyzes future patterns and trends of agricultural expansion by observing the expanded capabilities of 5G and modern technological developments. The chapter ends with Section 5.9, presenting opinionated insights regarding current development trends and their next phase applications.

5.1.1 Synopsis of conventional farming methods

Traditional agricultural approaches used for production rely mainly on automation as well as mono-cropping alongside synthetic pesticides. Cultural techniques boost land quality while preventing crop-infesting pests. Although modern farming practices boost environmental challenges, they create two main problems which damage soil quality and reduce biodiversity levels. The farming techniques have evolved into more sustainable methods of operation. The established agricultural system contains large-scale farming operations focused on boosting production and operational performance.

Farmers rely on synthetic fertilizers and insecticides, together with herbicides, to promote speedier growth while controlling pests with the conventional farming method. Auditory experts consider repeated planting of the same crop through mono-cropping to be detrimental because the practice depletes soil nutrients while making the land more vulnerable to pest attacks. Water infrastructure bears significant pressure, while irrigation systems, together with mechanization, help maintain high productivity rates [4]. Farmers select genetically modified organisms because their resistance to illnesses and pests provides popular advantages. The traditional farming methods, which use chemical run-off and natural biodiversity erosion as well as soil degradation, led to environmental hazards. Productivity is given priority in this system, frequently at the expense of long-term soil health and ecological balance [5]

The four stages of agriculture's evolution—Agriculture 1.0, 2.0, 3.0, and 4.0—are depicted in Figure 5.2. Manual labor was employed in agriculture 1.0, mechanization was introduced in agriculture 2.0, automation was introduced in agriculture 3.0, and precision agriculture was the main emphasis of agriculture 4.0. Every phase builds on the one before it, tackling issues and promoting a more robust and effective agricultural system.

5.1.2 Major obstacles in contemporary agriculture

Agriculture continues to be essential to both economic growth and global food security. However, there are several obstacles facing modern agriculture that jeopardize its production and sustainability. Major challenges such soil erosion, water scarcity, labor shortages, climate change, and economic hurdles are examined in this article. To lessen these difficulties and guarantee a robust agricultural system, the research highlights the necessity of creative approaches, legislative changes, and technology breakthroughs.

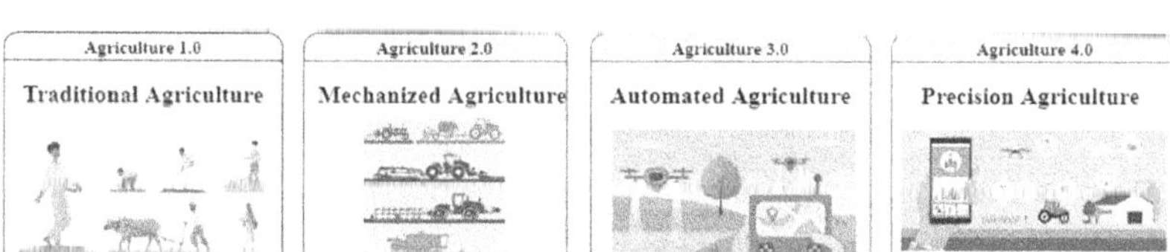

Figure 5.2 Revolution in agriculture sector.

Climate change crop yields, cattle, and fisheries are all greatly impacted by the unpredictable weather patterns, increased frequency of severe events, and warming global temperatures.

- **Soil Degradation:** Soil erosion, nutrient depletion, and the loss of arable land are the results of intensive farming, deforestation, and unsustainable agricultural methods.
- **Water Scarcity:** Approximately 70% of the freshwater used worldwide is for agricultural purposes. A significant obstacle to irrigation-dependent farming is the depletion of water supplies brought on by pollution, misuse, and climate change.
- **Labor Shortages:** The industry is now mostly dependent on mechanization due to a lack of workers brought on by urbanization and the aging of rural populations.
- **Economic and Policy Barriers:** Agriculture is not financially viable for many farmers due to problems including shifting market pricing, restricted finance availability, and insufficient government assistance.

The major obstacle remains economic and government policy specifications. Due to inferior markets and limited financial resources, as well as uncertain product prices, smallholder farmers experience weak competitive positions. Developing nations encounter considerable obstacles to obtaining technology-based solutions, which include precision farming and biotechnology. A comprehensive solution needs to be created to address wide-ranging agricultural problems, targeting equal availability of modern farming technology across all sectors [6]. A resilient food production system requires governmental, research-based, and agricultural community collaboration for solving these evaluation issues.

5.2 CHALLENGES AND BARRIERS

Climate change, soil erosion, water shortages, insect outbreaks, and ecological changes are some of the problems facing agriculture that jeopardize its sustainability and productivity. These include the consequences of climate change on crop output, animal health, and even agricultural resources, as well as the negative effects of over-cultivation [7], deforestation, and unsustainable farming practices on productivity and fecundity.

5.2.1 Infrastructure limitations

Poor infrastructure is a major constraint to the development and efficiency of the agricultural industry, including poor transportation networks, unstable electrical supply, and a shortage of storage facilities. The crops wilt due to this, and harvest losses and revenues are reduced. In addition, poor telecommunications infrastructure denies farmers access to crucial information. These challenges can only be overcome if rural communities have large investments in high-quality road development, energy connections, water supplies, and telecommunications [8].

5.2.2 Economic issues

Economic issues exacerbate these sustainable development and other lucrative prospects of the agriculture sector. Unstable market prices, high input costs, and a lack of capital exacerbate these issues [9]. In addition to facing loss shocks and obstacles to investing, subsistence farmers are also at risk from the state's little support. Actions like government support, price adjustments, and simple access to affordable financing are required to solve these issues, strengthen rural economies, and ensure the agricultural sector's financial viability.

5.2.3 Lack of education and training

Agricultural sustainability plus output fails because farmers lack proper education and training. Modern agricultural methods often fail to effectively handle two main problems, which include insect resistance alongside climate change effects. Farmers find it difficult to adapt their practices due to a lack of knowledge about farm regulations and market patterns and financial control procedures [10]. Widespread availability of agricultural education through extension services and training initiatives and accessible information will help farmers achieve better agricultural decisions, thus expanding the sector.

5.3 NEW DEVELOPMENTS IN AGRICULTURE TECHNOLOGY

Agricultural technology of today changes farming through its solutions for resource protection combined with food security maintenance and climate change management, while increasing yields and sustainability, while improving operational efficiency. These innovations base their developments on four essential components: Automation, data analytics, biotechnology, and smart farming methods [11].

5.3.1 Precision agriculture

Production management with modern equipment and IoT sensors, as well as GPS tracking and drone systems, allows farmers to achieve higher monitoring accuracy. Farmland monitoring enables farmers to track crop growth status, combined with soil and temperature measurements, along with moisture content. Methods using modern technology permit farmers to focus their usage of pesticides, fertilizers, and water more efficiently, minimizing wastage while improving production and reducing negative effects on the environment. Each field receives the exact field condition data through soil sensors that help farmers implement precise agricultural techniques utilizing fewer resources [12, 13].

5.3.2 Drones and aerial imaging

Drones with cameras and sensors inspect extensive agricultural land for monitoring purposes. The technology supplies time-sensitive information for crop conditions, together with water requirements while detecting insect presence through high-definition photos. By using these tools, farmers identify problems at an early stage so they can choose appropriate solutions that prevent serious consequences. Crop spraying stands as one of the drone applications that reduces labor expenses while ensuring higher precision levels.

5.3.3 Robotics and automation

Robotics systems execute all duties within agricultural practices, which include planting duties, along with weeding work, as well as harvesting processes and cow milk operations. These independent robots both enhance production speed and minimize human labor requirements on the farm. Modern agriculture benefits from robotic weeding devices that distinguish agricultural vegetation from unwanted growth so they can destroy undesirable species without harming crops, thus reducing pesticide requirements. The use of automated harvesters and tractors improves productivity by consuming reduced amounts of fuel and performing tasks in an advanced manner [14].

5.3.4 Applications of the IoT

Smart farms, where linked sensors continually monitor factors like soil moisture, temperature, and humidity, are being created using IoT devices. These gadgets exchange information with one another to enhance farming procedures and boost productivity. To save water and increase agricultural yields, smart irrigation systems, for instance, may modify watering schedules in real-time depending on soil and weather data. Figure 5.3 emphasizes how the IoT may improve sustainability and efficiency in agriculture. It demonstrates how drones offer aerial insights [15], water management systems track soil moisture, IoT-enabled machinery expedite processes, and soil monitoring sensors support accurate fertilization and management techniques. These solutions support environmental sustainability and data-driven decision-making.

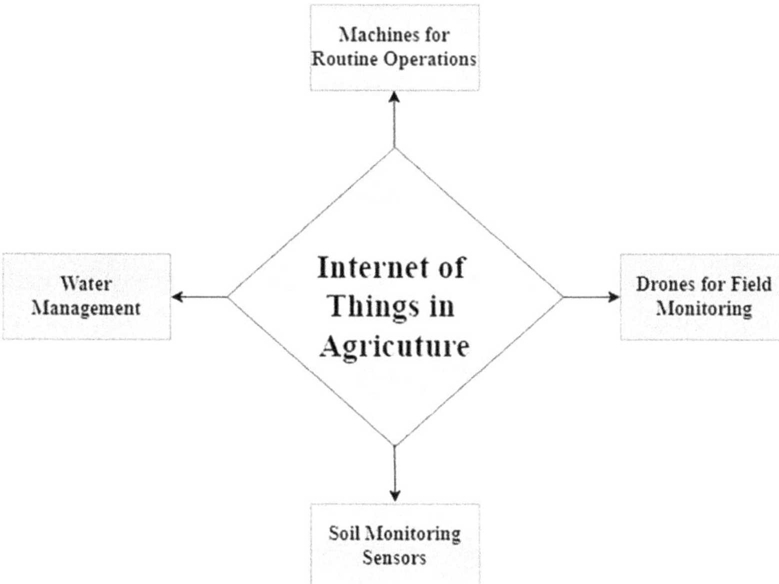

Figure 5.3 Overview of Agri-IoT.

5.3.5 Machine learning and AI

Large volumes of agricultural data are being analyzed by AI and machine learning algorithms to assist farmers in making data-driven decisions. AI systems can optimize irrigation schedules, forecast agriculture output, identify illnesses early, and even recommend when to grow or harvest crops. AI is assisting farmers in increasing output and lowering hazards since it can process and analyze data considerably more quickly than humans.

5.4 THE FUNCTION OF 5G TECHNOLOGY IN AGRICULTURE

5G technology is poised to revolutionize the agriculture sector by providing faster and reliable connections between devices, systems, and sensors. As a major goal of this technology, it is aimed at improving precision agriculture by enabling data collection and analysis in real time. 5G technology can help farmers by placing sensors in the fields, which can be used to inform them about the status of the crops, soil moisture content, and health of the soil [16]. Such information makes it possible to take timely actions such as adjusting irrigation schedules, applying fertilizers, or detecting insect outbreaks, which will improve agricultural productivity and resource utilization. Moreover, 5G-integrated drones and autonomous machinery will also be able to operate better by performing tasks such as drone seeding or harvesting with minimum human intervention and crop monitoring for imaging [17]. Another important aspect of 5G in agriculture is mechanizing IoT-based smart agriculture. Likewise, unprecedented farming modes that include cloud farming may emerge in the future due to 5G supporting IoT devices. IoT-enabled 5G devices can connect various elements of farming, for example, integrated farming systems to deliver services such as supply chains, logistics, silo, and equipment working together. For instance, delivery systems can monitor and manage optimal routes for transportation and smart silos can register the quality and levels of the grain. Also, 5G enables farmers to control and monitor their farms better than before, saving labor and increasing productivity [18].

Figure 5.4 demonstrates how 5G cellular networks have been used in contemporary agriculture to increase productivity, accuracy, and automation. Precision farming, planting by autonomous vehicles [19], watering, weeding, harvesting, animal monitoring, greenhouse management, and packing are important tasks. 5G guarantees a smooth connection between systems and devices, facilitating more intelligent, data-driven agricultural methods that increase output while preserving resources [20]. New methods of agriculture, such as the fourth agricultural revolution, are required due to the expanding world population and food production [19]. Big Data, AI, IoT, and precision farming are all used in this revolution to increase sustainability and efficiency. But the industry also has to contend with environmental and social issues, including the conflict between Russia and Ukraine, COVID-19, and climate change [18]. Smart farming, also known as Agri 4.0 [21], is a data-driven method of farming that makes use of real-time data from several sources to increase sustainability, production, and efficiency. These data are gathered and analyzed via wireless sensor networks ; radio-frequency identification and near-field communication are used to monitor agricultural assets, Bluetooth is utilized for interior farms, while ZigBee and LoRa are frequently found in outdoor settings. However, because of its high cost and energy consumption, Wi-Fi is less widely used [22]. Agriculture might undergo a transformation, thanks to smart farming, but its development is being hampered by a lack of affordable, dependable, and quick connections. High-speed, low-latency connectivity made possible by 5G technology can support applications such as the Massive IoT and make precision agricultural methods widely used. The difficulties and solutions of using 5G networks for smart farming are examined in this paper [23]. Figure 5.5 offers a framework for the "Sustainable Agriculture Revolution," emphasizing six environmentally friendly farming methods: precision agriculture, carbon farming, decarbonizing food supply chains, intercropping and crop rotation, beyond horticulture, and bio-fertilizers and microorganisms. By improving soil fertility and production, these tactics seek to build a resilient agricultural system that can satisfy food demands while protecting the environment [24].

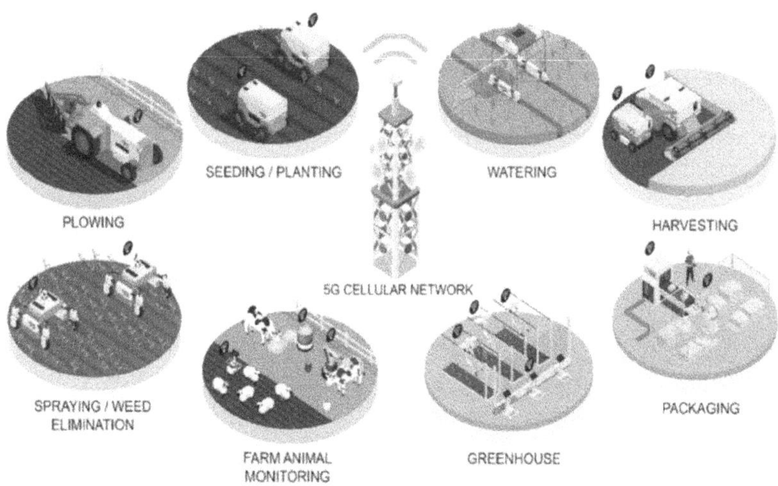

Figure 5.4 Precision agriculture using 5G networks as a communication.

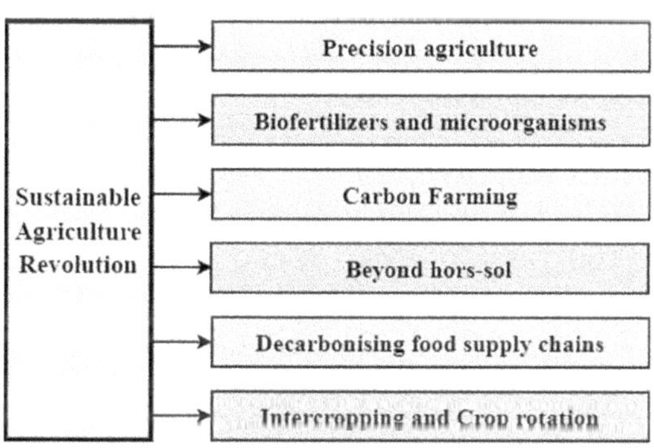

Figure 5.5 Sustainable agriculture revolution.

5.4.1 Improved data transmission and connectivity

Better connection and data transfer are essential for dependable, quick network communication. Efficiency may be increased by deploying edge computing, updating to Wi-Fi 6, upgrading to fiber optic cables and 5G networks, and upgrading infrastructure. Accurate transmission and reduced congestion are guaranteed via data compression methods, error correction systems, Quality of Service strategies, and load balancing [25]. The data flow operates under rules by traffic shaping and congestion management algorithms to alleviate network congestion by addressing bottlenecks. The combination of aggregation and caching results in better network performance, while encryption and virtual private network (VPN) implementation strengthen security parameters. Users can transfer their data securely, with complete privacy, through VPN technology. Those paths that data travels through can enhance reliability by letting traffic use multiple routes simultaneously, and Software-Defined Networking enables administrators to control network flow dynamically. The combination of multithreading and parallelism with optimal protocols like HTTP/2 or gRPC allows applications to achieve the highest speed and efficiency in data transmission. These methods, when used together, enhance performance and reliability and facilitate secure data transfers through networks.

5.4.2 Management and monitoring in real time

To guarantee the best possible performance, security, and dependability, real-time management entails keeping an eye on networks, applications, and systems. Sensors, agents, and APIs are used to gather data, analytics tools are used to process it, and dashboards are used to show system performance. Relevant staff get alerts, and problems are fixed with automatic reactions. Real-time monitoring is supported by tools like Datadog, Prometheus, Nagios, and New Relic, which offer detailed analysis and visual dashboards [26].

5.5 PRECISION FARMING

Figure 5.6 shows an ecosystem for smart farming that incorporates cutting-edge technologies like cloud computing, fog computing, and the IoT. It demonstrates the collection, processing, and use of data from diverse agricultural equipment to raise farm efficiency and productivity. IoT-enabled devices, a farmer's home network, a fog layer for local data processing, cloud computing for data analysis, and internet connectivity for data flow are all part of the farm's smart infrastructure. Farmers may implement precision agriculture, cut down on resource waste, and increase production, thanks to this integrated framework. Precision farming, also known as precision agriculture, uses technology to precisely monitor and manage animals and crops. It gathers data on crop health, weather, soil conditions, and irrigation requirements in real time using GPS, IoT sensors, drones, and data analytics.

Profitability is raised, yields are increased, and inputs are optimized with this knowledge. Smallholder farmers are finding it more and more accessible, and it promotes resource conservation and sustainable agriculture.

5.5.1 Meaning and significance

The farming method known as precision farming relies on GPS systems, IoT sensors, drones, and data analytics for handling agricultural activities with great accuracy. The implementation of precision farming technology enhances operational efficiency and sustainability while delivering better productivity

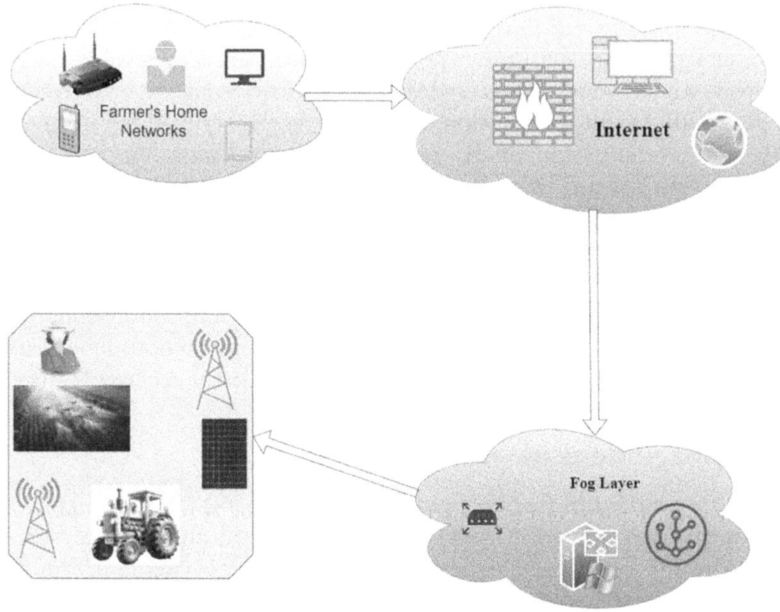

Figure 5.6 Developments in precision farming technology.

rates, decreasing waste output and production expenses, and minimizing environmental effects, resulting in increased crop outputs and maintaining ecological practices and resolving workforce deficits and climate stressors.

5.5.2 Precision farming methods assisted by 5G

The speedy and low-delay networks of 5G precision farming enable organizations to enhance agricultural sustainability alongside efficiency. The technology operates by processing current data from smart irrigation systems together with drones and autonomous machinery and IoT sensors and livestock monitoring equipment through AI and AI-powered systems. This technology installs better decision-making abilities through resource savings and supply chain traceability, and it eliminates rural connectivity gaps, thereby turning agriculture into a stronger and sustainable economic sector.

5.6 INTEGRATION OF 5G AND SUSTAINABLE PRACTICES

The combination of 5G technology capabilities with precision farming methods and resource optimization and waste reduction and sustainability drives complete agricultural transformation. IoT sensors supported by

5G-powered networks enable real-time crop and livestock monitoring, which leads to improved pest management and better fertilizer distribution and water irrigation decisions. The collaboration aims to resolve resource limitations and climate issues and dietary security matters so agriculture will eventually meet population needs while preserving environmental health. The combination of 5G technology and sustainable agricultural methods works toward improving productivity while also optimizing conservation and environmental protection. Farmers benefit from exact farming practices like minimal irrigation and fertilizer application, which become possible through IoT sensor connectivity, drone operation, and automated machinery functions. The 5G network supports efficient food waste reduction while resolving worldwide pollution and environmental deterioration problems through its smart supply chain and sustainable energy features.

5.6.1 Management of water and resources

Agricultural water and resource management serves the triple purposes of environmental conservation, sustainable production, and resource preservation. The use of water reaches maximum efficiency through the combination of rainfall collection, precise irrigation, smart storage, and drip irrigation techniques. Two sustainable agricultural methods include crop rotation systems and organic fertilization, while soil nutrient measurement equipment belongs to precision farming technology. Modern farming operations adopt renewable energy sources to decrease their dependence on non-renewable fossil fuels.

5.6.2 Monitoring of soil health

Being essential for sustainable agriculture, the monitoring of soil health enables resilient soils maintained with high richness. Helpful decisions can be made by farmers through their implementation of IoT devices, drones, and satellite pictures to monitor soil conditions immediately. Machine learning algorithms analyze erosion patterns to determine suitable solutions for prevention. Soil health gets enhanced through two sustainable techniques known as crop rotation and no-till farming. Grassland continues to be monitored to reduce environmental effects and achieve peak harvest quantity while sustaining farming operations indefinitely.

5.7 5G IMPLEMENTATION CASE STUDIES

The world witnesses a complete agricultural revolution attributable to 5G technology, which has driven significant progress in smart irrigation throughout Europe, Japan, Australia, the Netherlands, and India. Indian farmers benefit from IoT sensors that improve animal wellness, lower chemical usage, and maximize their water resources. The Japanese vertical farming methods optimize environmental factors to reduce energy expenses and enhance production outputs. Blockchain technology serves European supply chains to establish the ability to track products through their networks.

5.7.1 Using emerging technology to forecast

Agricultural Growth Forecasting: Agricultural development is being transformed by emerging technologies, which offer precise and effective resource management. Real-time information on crop health, soil conditions, and weather patterns is being made available by satellite photography, IoT sensors, drones, big data analytics, and blockchain technology. By tackling issues like climate change, resource scarcity, and unexpected weather, these technologies assist farmers in making well-informed decisions on planting dates, resource allocation, and harvesting tactics. An increasing global population is guaranteed food security through this transition to sustainable methods. In order to increase data processing efficiency and lower latency in real-time applications, Figure 5.7 depicts the example of convergence of edge computing with 5G technology. A 5G network infrastructure, an Edge Cloud, local edge cache units, a Multi-Access Edge Computing orchestrator, a Data Engine that uses AI, machine learning, deep learning, and pattern recognition, and a Resource Engine that uses software-defined networking, network slicing, network function virtualization, and self-organizing networks are all part of it. For industries like smart cities, healthcare, and agriculture, this architecture provides seamless communication, quick data processing, and effective resource management.

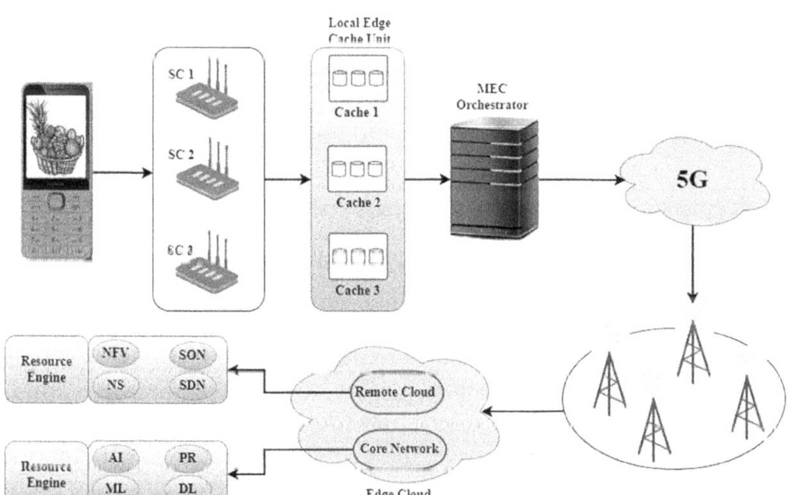

Figure 5.7 IoT and 5G architecture in general.

5.7.2 Globally successful agricultural innovations

Innovations in agriculture throughout the world are transforming farming methods and improving food security, sustainability, and efficiency. Food production, resource conservation, and environmental protection are being enhanced by technologies such as precision agriculture, vertical farming, CRISPR-Cas9 gene editing, aquaponics, hydroponics, smart irrigation, biological pest management, drones, and biodegradable packaging. These developments are changing agricultural methods, cutting waste, making the most use of available resources, and encouraging sustainable practices. These developments are changing how we work and live.

5.7.3 Expertise added

New innovations in agriculture produce more productive and sustainable farming methods. Multiple transformative technologies—such as drones, blockchain, regenerative agriculture, aquaponics, hydroponics, smart irrigation systems, precision farming, vertical farming, CRISPR gene editing, and biological pest control—collectively bring changes to the industry. Most innovations in our time drive productivity growth through optimal resource utilization, waste reduction, water conservation, and sustainable implementation.

5.8 PROSPECTS AND TRENDS FOR THE FUTURE

Agriculture's evolution will be directed by technological innovation and sustainable agricultural methods, which address environmental sustainability and worldwide food security demands. Increased deployment of AI, machine learning, automation, precision livestock farming, vertical and urban farming, and blockchain technology will result in agricultural systems that function more sustainably while also being better able to resist negative impacts. The future of agriculture will become efficient, sustainable, and resilient because of these developing trends.

5.8.1 5G's potential goes beyond

Present uses: 5G technology has the potential to completely transform social infrastructure and industries. Real-time communication between sensors, drones, and autonomous machinery is made possible by its low latency and fast connectivity, which improves resource management and precision farming. It can help with telemedicine, remote surgery, and better municipal administration. Additionally, it will improve the IoT and allow for driverless cars. Additionally, 5G will enable virtual and augmented reality, expanding access to immersive activities. Additionally, it will offer real-time analytics, increase security measures, and improve transaction efficiency. As 5G networks evolve, they will drive industry growth reliant on real-time data exchange.

5.8.2 Using emerging technology to forecast agricultural growth

A hybrid deep learning and feature extraction method for image-based classification, such as fruit categorization and plant disease detection, is shown in Figure 5.8. There are two primary steps in the process: handcrafted feature extraction, which breaks up photos into smaller subregions for more in-depth analysis, and deep learning, which extracts spatial and hierarchical characteristics from images. Dimensions are then reduced to 3,000 by using the mRMR feature selection on the merged feature collection. To improve accuracy and performance in image classification tasks, the improved feature set is then sent to a classifier for ultimate decision-making [13].

New technologies are transforming agricultural forecasting and giving farmers, academics, and policymakers precise information. Drones, AI, IoT sensors, and satellite imaging are important technologies. Better crop management is made possible by satellite imaging, which offers real-time information on weather patterns, soil conditions, and crop health. IoT sensors monitor environmental factors, and AI and machine learning analyze enormous volumes of data. Large farms can benefit greatly from the airborne data that drones gather on crop health and environmental factors. The agricultural supply chain is transparent and traceable, thanks to blockchain technology. These technologies provide greater production, resilience in the face of global problems, and sustainable management of agricultural resources.

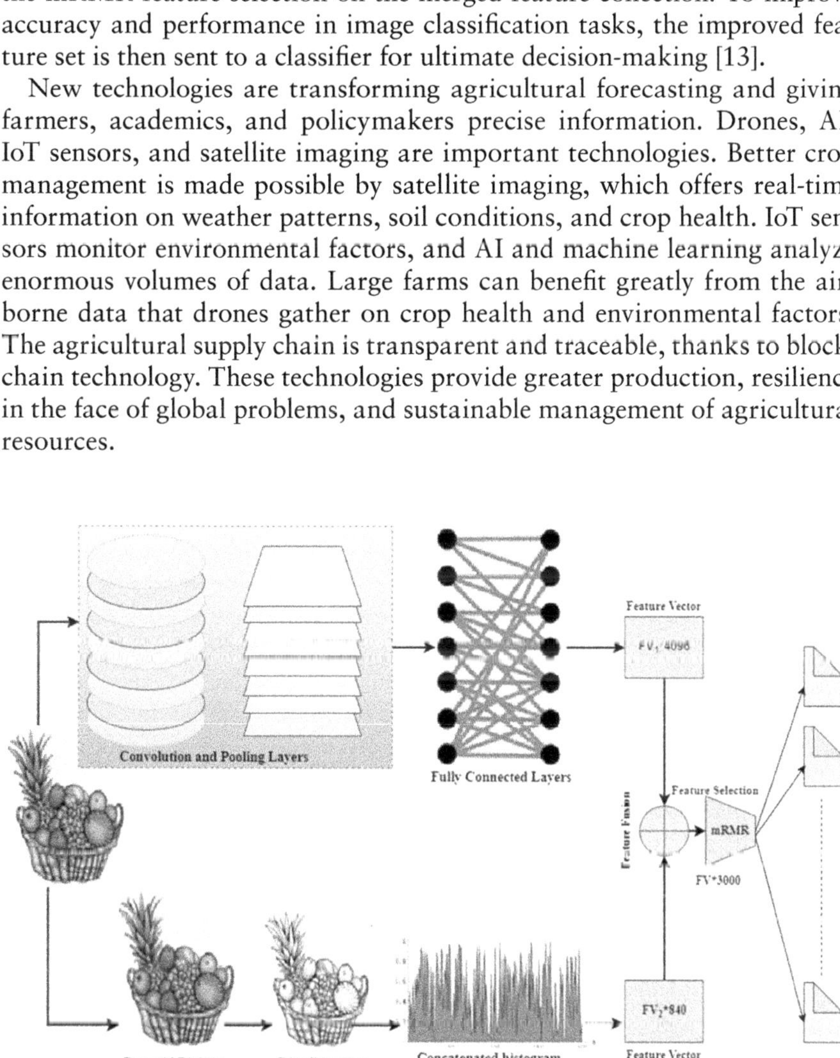

Figure 5.8 A Deep Learning based approach for plant disease categorization and disease classification.

5.9 CONCLUSION AND FUTURE SCOPE

The solution to agricultural industry problems depends on the combination of 5G technology and aerial imaging while utilizing AI, IoT, and machine learning. Modern farming methodologies have become possible through these technologies that use networking to solve issues regarding marginalization and isolation of farms located on sloping terrain. 5G technology enables improved chemical and fertilizer management and better irrigation systems through advanced networking capabilities. These advancements create possibilities for stakeholders to establish worldwide food security along with promoting sustainability practices while achieving a significant increase in agricultural yields. This chapter uses empirical examples with optimistic indications to show how these technologies have the capability of disrupting the agricultural value chain entirely and proving their sustainability as global-scale positive changes. The chapter evaluates how 5G technology, AI, IoT, and machine learning can transform the agricultural sector. The research shows how smart irrigation technology combines well with improved demand analytics and green farming procedures. The chapter shows practical examples from real-life situations that explain both the operational effects of these technologies in addition to their long-term beneficial effects on worldwide agricultural value chain operations. The implementation of 5G combined with AI technology, IoT elements, and machine learning abilities into agriculture needs combined efforts from farmers through government support with technological providers, academic research, and consumer participation. Governments need to back the adoption through three main actions: providing necessary infrastructure, financial backing, and relevant policy frameworks. Technology providers must provide guaranteed access to their systems, and academic institutions, along with researchers, must seek innovative solutions. Customers should demand complete openness from manufacturers regarding their production methods for food items. Various stakeholders need to collaborate because this collaboration will enhance food security and environmental protection and be quick.

REFERENCES

[1] M. A. Khan, A. Khan, M. Abuibaid, and J. S. Huang, "Harnessing 5g networks for enhanced precision agriculture: Challenges and potential solutions," in *2023 International Conference on Smart Applications, Communications and Networking (SmartNets)*, Istanbul, Turkiye, pp. 1–6, IEEE, 2023.

[2] B. Wang, "Smart farming using the big data-driven approach for sustainable agriculture with IoT: Deep learning techniques," *Scalable Computing: Practice and Experience*, vol. 25, no. 2, pp. 675–682, 2024.

[3] Y. Tang, S. Dananjayan, C. Hou, Q. Guo, S. Luo, and Y. He, "A survey on the 5g network and its impact on agriculture: Challenges and opportunities," *Computers and Electronics in Agriculture*, vol. 180, p. 105895, 2021.

[4] A. Nag, A. Das, N. Chand, and N. Roy, "Sustainable agriculture: A critical analysis of internet of things-based solutions," in *Intelligent Systems and Industrial Internet of Things for Sustainable Development*, pp. 118–138, Chapman and Hall/CRC, 2024.
[5] E. Tsormpatsidis, M. Ordidge, R. G. Henbest, A. Wagstaffe, N. H. Battey, and P. Hadley, "Harvesting fruit of equivalent chronological age and fruit position shows individual effects of UV radiation on aspects of the strawberry ripening process," *Environmental and Experimental Botany*, vol. 74, pp. 178–185, 2011.
[6] M. Louta, K. Banti, and I. Karampelia, "Emerging technologies for sustainable agriculture: The power of humans and the way ahead," *IEEE Access*, vol. 12, pp. 98492–98529, 2024.
[7] S. Ji and A. K. Mishra, "5g security issues challenges and solutions against DDoS attacks: A survey," in *2024 2nd International Conference on Disruptive Technologies (ICDT)*, Greater Noida, India, pp. 1422–1427, IEEE, 2024.
[8] S. Ji and A. K. Mishra, "5g network implementation: A survey on security issues, challenges, and future directions," in *Developments Towards Next Generation Intelligent Systems for Sustainable Development*, pp. 62–88, IGI Global, 2024.
[9] V. Jayaswal, R. Rani, and J. Kaur, "A deep learning-based efficient image captioning approach for Hindi language," in *Developments Towards Next Generation Intelligent Systems for Sustainable Development*, pp. 225–246, IGI Global, 2024.
[10] M. S. Ali, A. M. M. Raafee, and A. S. M. Mia, "Smart farming using ai towards Bangladeshi agriculture," *Journal of FST*, vol. 1, no. 01, p. 189, 2022.
[11] M. Amiri-Zarandi, R. Dara, E. Duncan, and E. Fraser, "Big data privacy in smart farming: A review," *Sustainability*, vol.14, no. 15, p. 9120, 2022.
[12] Y. Tang, S. Dananjayan, C. Hou, Q. Guo, S. Luo, and Y. He, "A survey on the 5g network and its impact on agriculture: Challenges and opportunities," *Computers and Electronics in Agriculture*, vol. 180, p. 105895, 2021.
[13] M. Javaid, A. Haleem, R. P. Singh, and R. Suman, "Enhancing smart farming through the applications of agriculture 4.0 technologies," *International Journal of Intelligent Networks*, vol. 3, pp. 150–164, 2022.
[14] A. Lytos, T. Lagkas, P. Sarigiannidis, M. Zervakis, and G. Livanos, "Towards smart farming: Systems, frameworks and exploitation of multiple sources," *Computer Networks*, vol. 172, p. 107147, 2020.
[15] C. N. Verdouw, J. Wolfert, A. Beulens, and A. Rialland, "Virtualization of food supply chains with the internet of things," *Journal of Food Engineering*, vol. 176, pp. 128–136, 2016.
[16] F. Zhang, Y. Zhang, W. Lu, Y. Gao, Y. Gong, and J. Cao, "6g-enabled smart agriculture: A review and prospect," *Electronics*, vol. 11, no. 18, p. 2845, 2022.
[17] Z. A. S. Aqeel-ur Rehman, N. A. Shaikh, and N. Islam, "An integrated framework to develop context-aware sensor grid for agriculture," *Australian Journal of Basic and Applied Sciences*, vol. 4, no. 5, pp. 922–931, 2010.
[18] A. Khan, J.-P. Li, R. A. Shaikh, and I. Khan, "Vision based classification of fresh fruits using fuzzy logic," in *2016 3rd International Conference on Computing for Sustainable Global Development (INDIACom)*, New Delhi, India, pp. 3932–3936, IEEE, 2016.

[19] G. Obi Reddy, B. Dwivedi, and G. Ravindra Chary, "Applications of geospatial and big data technologies in smart farming," in *Smart Agriculture for Developing Nations: Status, Perspectives and Challenges*, pp. 15–31, Springer, 2023.
[20] T. A. Shaikh, W. A. Mir, T. Rasool, and S. Sofi, "Machine learning for smart agriculture and precision farming: Towards making the fields talk," *Archives of Computational Methods in Engineering*, vol. 29, no. 7, pp. 4557–4597, 2022.
[21] S. Yadav, A. Kaushik, M. Sharma, and S. Sharma, "Disruptive technologies in smart farming: An expanded view with sentiment analysis," *AgriEngineering*, vol. 4, no. 2, pp. 424–460, 2022.
[22] V. Balafas, E. Karantoumanis, M. Louta, and N. Ploskas, "Machine learning and deep learning for plant disease classification and detection," *IEEE Access*, vol. 11, pp. 114352–114377, 2023.
[23] A. Chatzisavvas, M. Louta, and M. Dasygenis, "Implementation of agricultural path planning with unmanned ground vehicles (UGV) based on enhanced A* algorithm," in *2023 12th International Conference on Modern Circuits and Systems Technologies (MOCAST)*, Athens, Greece, pp. 1–4, IEEE, 2023.
[24] A. Chatzisavvas, M. Dasygenis, and M. Louta, "Autonomous unmanned ground vehicle in precision agriculture: The VELOS project," in *2022 7th South-East Europe Design Automation, Computer Engineering, Computer Networks and Social Media Conference (SEEDA-CECNSM)*, Ioannina, Greece, pp. 1–4, IEEE, 2022.
[25] J. Du, G. Han, C. Lin, and M. Martinez-Garcia, "ITrust: An anomaly-resilient trust model based on isolation forest for underwater acoustic sensor networks," *IEEE Transactions on Mobile Computing*, vol. 21, no. 5, pp. 1684–1696, 2020.
[26] I. M. Nasir, A. Bibi, J. H. Shah, M. A. Khan, M. Sharif, K. Iqbal, Y. Nam, and S. Kadry, "Deep learning-based classification of fruit diseases: An application for precision agriculture," *Computers, Materials and Continua*, vol. 66, no. 2, pp. 1949–1962, 2021.

Chapter 6

Exploring the role of machine learning and blockchain in the context of IoT

Toward a smart agriculture system

Sandeep Bhatia, Zainul Abdin Jaffery, Shabana Mehfuz, and H. S. Ganesha

6.1 INTRODUCTION

India's farming industry is essential to the country's economic development and employment creation. In India, 49% of the population makes their livelihood from agriculture. Villagers in India benefit greatly from agriculture, which also boosts the country's economy and offers large-scale development potential. 38.4% of the world's land area is used for agriculture; 54% of India's land area is utilized for this purpose, and 42% of India's land is irrigated; India accounts for 2.4% of the world's total land area, according to the analytical study on the status of agriculture in India [1]. The agriculture sector employs around half of all workers in the country. On the other hand, at 2015–2016 prices, its GDP share is 17.5%. The agricultural sector's GDP share fell from over 50% in the 1950s to 15.4% in 2015–2016 (at constant prices), while the industrial and services sectors have been increasingly important to the economy's growth in recent decades.

India ranks among the world's leading producers of several crops, such as cotton, sugarcane, wheat, rice, and pulses. The amount of food grains it produces has increased yearly. It is the world's top producer of milk and comes in second for fruits and vegetables. In terms of percentage, India led the globe in 2013 with 25% of pulses, 22% rice, and 13% wheat produced.

Nine billion people will inhabit the earth by 2050 (World Bank). To keep up with demand, 70% more food must be produced. Therefore, ensuring food sustainability is a highly challenging issue. Profits and productivity are being lost by using outdated methods. Conventional agricultural practices are also being impacted by the environment, according to the Nabard Report [2]. Traditional farming methods thus confront several difficulties. Precision agriculture will play a major role in the current farm revolution. It will cut labor, input, and time in a sustainable manner. It will maximize production as well as profits. It will ensure durability and reduce the impact of the environment. Precision farming is a kind of agricultural management that makes use of technology to track, measure, and analyze the unique needs of every field and crop. Most nations primarily rely on agriculture as their primary source of revenue. Agriculture is the main sector in developing

DOI: 10.1201/9781003494782-6

countries. It ensures the food security of a country and produces resources for industry [3]. In the pursuit of efficient and sustainable farming, the shift from conventional farming methods to smart agriculture powered by IoT represents a critical turning point. Agriculture has always been at the forefront of human progress, developing constantly to improve quality and production to satisfy the expanding demands of the world's population. The introduction of the Internet of Things (IoT) into agriculture is a significant advancement that aims to transform conventional farming methods [4]. IoT technology makes it easier to monitor important agricultural factors, which results in smarter, more productive farming methods that lower labor expenses, increase work intensity, and lessen the need for manual agriculturalist intervention through automation [5]. The primary activity of growing crops and raising animals for both economic and human use is known as agriculture. It has supported human civilization for thousands of years, giving billions of people throughout the planet food, raw resources, and a means of subsistence [6]. Scarcity of water is a major problem in many places worldwide. Water shortages cannot be addressed by traditional irrigation methods, which use more water than necessary because agriculture is frequently the sector that uses the most water. The process of creating a water allocation plan for agricultural use that is both efficient and reasonably priced requires the application of optimization models. The world's growing population is driving up food consumption, which is having an adverse effect on the environment and seriously taxing agricultural output. There is a significant impact on the output of agriculture. Many obstacles need to be overcome to boost productivity, such as the annual reduction in land area, the gradual decline in soil fertility, farmers' lack of knowledge about agricultural technology, irrigation issues, and financial constraints [7,8]. Ineffective weather forecasting techniques and unforeseen weather patterns are other factors that reduce agricultural production [9]. The fourth generation of agricultural technology focuses on four key areas: increasing productivity, judicious use of resources, mitigating the effects of climate change, and decreasing food waste. The IoT refers to a collection of networked devices or equipment that communicate and collect data on their own over a wireless network. It is used for many different things, such as asset identification, security, automated construction, and medical. The IoT-based data collection system periodically collects data about the physical surroundings. Big data analytics, video surveillance, remote monitoring, and other applications can benefit from these systems [10]. Numerous techniques have been used to apply the network to performance optimization. It is a cloud-based technology that works with individuals worldwide to do various tasks that are assigned to it. Devices such as PCs, tablets, smartphones, and other computing platforms are connected to the Internet using a router. After that, the data are transmitted over Bluetooth, Wi-Fi, and other similar technologies [9,10]. Connecting devices that allow for device-to-device communication is mostly the task of the IoT. IoT designs will facilitate online communication and device control. The IoT uses a variety of sensors that are placed in agricultural regions to transmit data, including plant images for monitoring soil health and crop, humidity, and temperature sensors. When it detects it, the gadget sends information over the internet [11]. Precision agriculture is a knowledge-driven approach to farm management that considers, evaluates, and analyzes the needs of certain regions and crops. The use of data and information technology in sophisticated agricultural operations is known as intelligent farming. "Smart farming," which seeks to transform the way food is produced and maintained, is the newest buzzword in agriculture.

One way to conceptualize smart farming is the advancement of precision agriculture. A few of the contemporary technologies used in smart farming include machine learning (ML), big data, cloud computing, and wireless sensor networks, or WSNs. The objective of smart farming is to reduce production costs while increasing output and improving product quality [12,13]. The technologically inventive force known as ML has ushered in a new era of possibilities. ML is the engine behind many breakthroughs, from customized suggestions to result prediction, and it promises a future in which computers will be able to learn from the data they encounter in addition to understanding it. The effects of climate change have been blamed for the significant drop in agricultural performance that most of India's crops have seen over the past 20 years. Anticipating agricultural output before the actual harvest becomes essential to provide farmers and policymakers with useful information for efficient marketing and storage decisions. As a revolutionary tool, ML enables farmers and other agricultural stakeholders to make well-informed decisions, use resources wisely, and eventually increase crop yields [14]. The process of developing a prediction system that works requires the use of several different algorithms and their linkages. The agriculture sector is expanding its horizons and embracing the promise of data-driven insights to overcome obstacles by utilizing big data analytics approaches. The acceptance of undesirable goods and processes has caused the agriculture sector to gradually deteriorate, which has increased the demand for novel ways. The transition to synthetic and hybrid products, along with ignorance of the best ways to cultivate, has changed seasonal climates, led to an unhealthy lifestyle, and depleted vital resources like soil, water, and air, all of which have finally resulted in a shortage of food [15]. In the face of these obstacles, policies that support India's agricultural economy must be implemented. Improving crop quality and productivity is a primary goal, and ML—a branch of artificial intelligence (AI)—becomes an effective ally in this attempt. We want to provide farmers with accurate advice by utilizing data-driven insights, which will enhance crop planning and boost agricultural output [16]. Blockchain is a decentralized peer-to-peer (p2p) network used to track all transactions on the blockchain network. They lack a single point of contact, which means that a group of nodes manages the network instead of a single entity or individual. This adds an extra layer of security to improve transaction security within the network. Records are duplicated at the system level at every node. Each block holding the transaction's information is added by the node, which must first confirm the transaction's validity. Many different sectors utilize blockchain technology. Since the blockchain uses a decentralized storage system technology, it benefits data security and authenticity in several ways and contributes a lot toward the agriculture sector [17]. Smart farming uses blockchain technology to improve supply chain management, forecast agricultural yields, and guarantee the safety of the food supply chain. It creates a safe means of disseminating data and permits information tracking along the supply chain. Blockchain technology is utilized in contract farming to establish smart contracts that link farmers with supply firms and landowners. Farmers may now lease land, buy seeds and fertilizer with little to no capital outlay, and avoid the need to qualify for loans. Furthermore, smart cattle farming makes use of blockchain technology to enhance operational efficiency, ecological sustainability, and economic viability [18,19].

Blockchain and ML can be combined with IoT to enable the construction of a smart agricultural framework, which is a new revolution. The goal of this combination is to improve security within the scope of IoT-based agriculture. Using sets of sensors, such as soil moisture, pH, rainfall detectors, temperature sensors, and crop disease detection, ML for smart agriculture framework in the context of the IoT can be used in precision agriculture, crop recommendation systems, water and waste management, livestock monitoring, farm management, pest management, and predictive analytics to predict suitable weather conditions for efficient crop management.

The introduction to blockchain and ML in the context of the IoT is covered in Section 6.1 of this chapter. Section 6.2 contains the literature survey. A summary of several ML algorithms for optimizing smart agriculture systems for the study of various agriculture potential risks in the blockchain and IoT-enabled smart agriculture framework (IESAF) will be presented in Section 6.3. Section 6.4 covers the integration of blockchain and ML approaches for the IESAF.

Using a specific set of sensors (such as temperature, pH, and moisture sensors) and communication technologies (such as Lora WAN, ZigBee, Wi-Fi, and Bluetooth), the proposed architecture of ML and blockchain in the context of IoT for a smart agriculture system sends agricultural data collection results to a remote location for optimization, which is discussed in Section 6.5 with an explanation of various components used.

Farmers now have a means to boost crop output with this technique. Section 6.6 discusses integration concerns such as security, scalability, and interoperability to provide sustainable, more effective, and favorable settings for precision agriculture. Section 6.7 presents future research directions related to smart agriculture. Section 6.8 includes a case study on smart agriculture and a perspective on how to boost production using the suggested approach. Section 6.9 also contains the chapter's conclusion.

The references used to prepare the chapter were included at the end. Our suggested methodology helps farmers choose the best course of action that boosts crop development.

6.2 LITERATURE REVIEW

The integration of blockchain technology and ML in IoT-based smart agriculture is a promising development, with potential to enhance security and efficiency [20]. ML can be applied in various aspects of smart agriculture, including precision agriculture, crop recommendation, and pest management [20,21]. The use of IoT and ML in smart farming can lead to increased productivity and reduced wastage [22]. The combination of IoT and blockchain technology in smart farming can also improve data security and transparency [23]. In [24], authors present a comprehensive overview of an IoT-based smart farming system employing ML techniques to optimize plant growth and support farmers in their activities. In contrast, Chaganti et al. [25] propose a blockchain-based cloud-enabled security monitoring framework for smart agriculture, addressing the critical issue of security in IoT deployments. Gayatri et al. [26] explore the intersection of IoT and cloud computing in agriculture to provide smart solutions for farmers, emphasizing the importance of leveraging emerging technologies to address challenges in agrarian output. Kaushal [27] discusses the integration of IoT and blockchain technology in smart agriculture. The paper provides an overview of IoT applications in agriculture, emphasizing the need for leveraging emerging technologies to address agricultural challenges, thus contributing to ongoing research efforts in smart agriculture. Singh and Singh [28] conduct a thorough review of the integration of blockchain, IoT, and AI in the agriculture and healthcare sectors, emphasizing their roles in enhancing security, transparency, and efficiency. Torky and Hassanein [29] delve into the integration of blockchain and IoT, specifically in precision agriculture. Pranto et al. [30] address the need for modernization in agriculture through the integration of blockchain, smart contracts, and IoT, proposing a system to automate processes, establish trust among stakeholders, and ensure transparency and fairness in agricultural transactions. Padiya et al. [31] highlight the transformative impact of these technologies in optimizing crop yield, resource utilization, and farm management, underscoring their significance in revolutionizing the agricultural sector. Gupta, Malhotra, and Kumar [32] present a bibliometric review of 42 articles, focusing on the integration of ML algorithms within IoT networks for smart agriculture applications. Ahmed et al. [33], on the other hand, talk about climate-smart agriculture (CSA) using smart techniques, blockchain, and the IoT. They provide an overview of IoT, AI, ML, and blockchain technologies and propose approaches for integrating them into CSA systems. Maduranga and Abeysekera [34] review the applications of ML in IoT-based agriculture and smart farming, focusing on data-driven farm management techniques to increase agricultural yields and resource efficiency. Biswas et al. [35] propose a blockchain-based smart agriculture with the IoT (BIoT) model, emphasizing the transformative potential of blockchain technology in agriculture. Friha et al. [36] provide a comprehensive survey of emerging technologies for IoT-based smart agriculture, covering various aspects such as unmanned aerial vehicles, wireless technologies, and blockchain for supply chain management. Kumar and Dwivedi [37] propose a methodology for transforming traditional agriculture into smart farming using blockchain and IoT technologies.

They emphasize the importance of data collection, recording, and verification in ensuring food safety and security throughout the agricultural supply chain. Agricultural land is being monitored with little human involvement using tools such as IoT-WSN-based systems. Numerous wireless protocols are now in use, such as Wi-MAX, LoRa WAN, RFID, MQTT, Zigbee, Sigfox, and BLE (Bluetooth low energy) [38,39,40]. Over the next 10 years, the farming sector will depend heavily on the superfast 5G connection to improve crop yields and quality with the least amount of labor. Precision and intelligent farming may help farmers become more informed and productive. With the advent of 5G, the nature of employment in farming and agriculture will change dramatically. Flexible and effective solutions provided by the IoT-based WSN of the 5G network can help to enable smart farming. They will make it easier for various autonomous agricultural equipment to operate automatically throughout crop cultivation phases, including planting, ploughing, and management [41]. The ultimate objective is to enable the creation of unmanned farms while achieving safe, dependable, environmentally friendly, and energy-conserving operations [42]. An architecture for an IoT-based Wireless Sensor Network (WSN) that aims to improve security and energy efficiency in smart agricultural applications is presented ref [43]. The integration of IoT technology into cybersecurity measures in the healthcare industry is thoroughly analyzed in the study. The authors list and go over the cybersecurity issues that the spread of IoT devices in healthcare settings has brought about [44]. To improve crop growth efficiency, a smart farming system utilizing IoT technology is presented. The authors examine a range of IoT technologies that are pertinent to agriculture, including weather stations, drone-based aerial imaging systems, and soil moisture sensors [45]. Research on the use of IoT technology for smart farming implementation is presented. The authors outline the suggested system architecture for integrating IoT technologies with smart farming [46]. The article uses use cases and case studies to show how IoT technology is being used in smart farming in the real world

[47]. A conceptual framework for IoT dependability is presented in this chapter, along with definitions of important terms and elements, including fault tolerance, resilience, and system availability. It addresses how security, scalability, performance, and dependability interact in IoT deployments and emphasizes the necessity of comprehensive strategies to guarantee stable and trustworthy IoT systems [48]. An extensive analysis of the use of IoT technology in smart farming is presented in the study. IoT sensors and devices, data analytics methods, precision farming techniques, and case studies of IoT installations in agricultural environments are a few examples of these subjects [49]. The creation and study of an IoT-based smart agricultural system that can support heterogeneous nodes is presented in the work. The article addresses how IoT technology supports smart agricultural techniques, including environmental sensing, crop monitoring, and precision farming [50]. The integration of drone technology with intelligent irrigation systems for agricultural purposes is examined in the study. The article addresses drone technology's use in agriculture, emphasizing its use in precision farming, insect identification, and crop monitoring [51]. The application of precision agricultural methods made possible by IoT technology is examined in the study [52].

Table 6.1 shows the similar works in this field represented in the tabular form.

Table 6.1 Similar work in this field

Author's information	*Objective of the paper*	*Methodology*	*Gaps identified*
Ahmad and Nabi [3]	Explore the integration of AI, IoT, and ML in Agriculture 5.0.	Literature review and framework analysis.	Limited real-world implementation examples and performance evaluations.
Dhawale [4]	Review IoT-based smart agriculture systems.	Analysis of existing IoT technologies and their applications.	Lack of cost-effective solutions for small-scale farmers.
Amandeep et al. [5]	Develop a smart farming system using IoT.	Implementation of IoT devices for real-time monitoring.	Scalability and energy efficiency concerns.
Zheng et al. [6]	Develop a mobile-based smart farming service system.	System design and implementation using mobile technologies	Limited scope for modern IoT advancements.

(*Continued*)

Table 6.1 (Continued) Similar work in this field

Author's information	*Objective of the paper*	*Methodology*	*Gaps identified*
Telkomnika [7]	Propose a smart agriculture management system using IoT.	Prototype development with IoT integration.	Insufficient focus on data security and privacy.
Baseca et al. [8]	Propose an IoT-oriented system for agriculture monitoring.	Design and deployment of IoT-based monitoring systems.	Lack of integration with advanced analytics.
Qazi et al. [9]	Review AI-enabled and IoT-equipped smart agriculture systems.	Comprehensive survey of IoT and AI technologies.	Challenges in interoperability and data standardization.
Subhashini et al. [12]	Develop a wireless plant sensor for smart farming.	IoT-based wireless sensor network design.	Limited scalability and high maintenance costs.
Quy et al. [13]	Explore IoT-enabled smart agriculture architecture and applications.	Architectural analysis and case studies.	Insufficient exploration of IoT's economic impact on farmers.
Dahane et al. [14]	Propose an IoT-based smart farming system using ML.	System design with ML algorithms.	Limited real-world deployment and validation.
Ahmed et al. [15]	Examine IoT's role in smart precision agriculture for rural areas.	Case studies and implementation examples.	Lack of accessibility for low-resource settings.
Bhanu et al. [16]	Propose an IoT-based system for enhanced irrigation.	System design with IoT-enabled irrigation techniques.	Limited adaptability to diverse climatic conditions.
Awan [23]	Investigate the role of IoT with blockchain in smart farming.	Conceptual analysis of IoT-blockchain integration.	Lack of field-level implementation and scalability analysis.
Chaganti et al. [25]	Implement blockchain-enabled IoT security monitoring for smart agriculture.	Blockchain and IoT system integration.	High computational costs and energy demands.
Gayatri et al. [26]	Provide smart agricultural solutions using IoT.	Development of IoT-enabled tools for farmers.	Limited focus on long-term sustainability.
Friha et al. [36]	Survey IoT technologies for smart agriculture.	Comprehensive review of emerging IoT technologies.	Lack of focus on region-specific challenges and solutions.
Yoon et al. [45]	Implement a smart farm using IoT technology.	IoT-based system design and field trials.	Limited user-centric customization options.
Akella et al. [52]	Review blockchain adoption barriers and enablers in agriculture.	Systematic literature review.	Insufficient empirical studies and real-world validation.

6.3 EXISTING PROBLEMS IN THIS RESEARCH AREA

Many difficulties emerge when integrating blockchain technology and ML with IoT-based smart agriculture systems. The difficulty of interoperability and data integration is one of the main problems. Large volumes of heterogeneous data are generated by IoT devices in agriculture; ML methods are required to process and interpret these data. On the other hand, it may be more difficult to ensure smooth communication between gadgets, algorithms, and blockchain systems in the absence of defined data formats and protocols. This could result in inefficiencies, data fragmentation, and a decline in the system's overall efficacy [53].

The energy consumption and scalability of these systems provide another significant concern. Substantial computational resources are needed for ML models, particularly for deep learning algorithms. On the other hand, blockchain, especially when using energy-intensive consensus techniques like proof of work, requires far more energy. The electricity supplies that are available in agricultural settings may be strained by these requirements, especially in isolated or off-grid places. Furthermore, there is a concern of blockchain scalability because it cannot support large-scale agricultural operations' real-time processing requirements, which may limit its usefulness and acceptance in the industry [54].

Table 6.2 shows a summary of the main objectives, applications, difficulties, and future directions of the study and provides the context for smart agricultural frameworks that are made possible by blockchain, IoT, and ML technologies.

Table 6.2 Summary of the study in a tabular form

Aspect	*Description*
Objective	To develop an IoT-enabled smart agriculture framework by combining blockchain technology with ML to enhance security and efficiency in agricultural operations.
Applications of ML in IoT-based Smart Agriculture [5]	• Precision Agriculture • Crop Recommendation Systems • Water and Waste Management • Livestock Monitoring • Farm Management • Pest Management • Predictive Analytics for Weather Conditions • Crop Disease Detection
Specific Sensors Used [6]	• Soil Moisture Sensors • pH Sensors • Rainfall Detectors • Temperature Sensors

(*Continued*)

Table 6.2 (Continued) Summary of the study in a tabular form

Aspect	*Description*
Applications of Blockchain in IoT-based Smart Agriculture [7]	• Secure Data Transmission • Supply Chain Management • Smart Contract Systems
ML Models Discussed [8]	• SVM • CNN • LSTM • GA • RA • DTs • RF • Clustering • KNN • GNB
Integration Challenges [9]	• Security • Scalability • Interoperability
Future Research Directions [15]	Exploration of sustainable, more efficient, and favorable conditions for precision agriculture through advanced integration of IoT, blockchain, and ML technologies.
Case Study Focus [18]	Discussion of a case study related to smart agriculture with a proposed framework designed to assist farmers in finding optimal solutions, leading to increased crop growth and productivity.

6.4 SUMMARY OF SEVERAL ML ALGORITHMS FOR OPTIMIZING SMART AGRICULTURE SYSTEMS

For the analysis of various agriculture threats in blockchain and IESAF, there are several ML models available, such as Support Vector Machine (SVM), Convolutional Neural Network (CNN), Long Short-Term Memory (LSTM), Genetic Algorithm (GA), Regression Algorithm (RA), Decision Trees (DT), Random Forest (RF), Clustering and K-Nearest Neighbors (KNN), Gaussian Naïve Bayes (GNB) [55, 56].

6.4.1 Support vector machine

By extending the border line between training set classes in the hyper plane, SVMs classify data. Hyperplanes are expressed using the following formula:

$$f(x) = a^T x + c \tag{6.1}$$

In this formula, a denotes the dimensional coefficient, and c denotes the offset in this case.

6.4.2 Convolutional neural network

Digital images are binary representations of visual data, where color and brightness values are assigned to individual pixels arranged in a grid structure. CNN excels in analyzinggrid based Input data such as images. CNNs effectively extract spatial and hierarchical features from this grid-based representation.

6.4.3 Long short-term memory

In smart agriculture, LSTM models can be used for crop prediction, disease detection, and weather forecasting. The model can examine past environmental data and be used for crop yield prediction.

6.4.4 Genetic algorithm

GA works on the process of evolution and natural selection. This model can be used in smart agriculture to enhance productivity, longevity of soil, and promote sustainable agriculture.

6.4.5 Regression Algorithm

To predict values based on input data, RAs is used. In smart agriculture, this method can be used to maximize crop yield and enhance efficiency of the system.

6.4.6 Decision trees

It functions on the greedy principle and makes use of a collection of categorization criteria. Alternatively, this approach explains why noisy data cannot be handled by the technique, even if it has great categorization accuracy. Gain is the primary factor that determines the root node parameter of the DT. C4.5 is referred to as a fuzzy statistical classifier because to the fact that it produces DTs that may be used for classification [56].

$$E(S) = \sum_{i=1}^{c} Pi(1 - \log 2) \tag{6.2}$$

6.4.7 Random forest

RF is a variant of the DT technique. DTs contain significant volatility and little bias; hence, the variance component of the model is decreased by averaging them. By averaging the forecast, samples that are unknown can be created.

$$I = \frac{1}{N} \sum_{n=1}^{N} f(x) \tag{6.3}$$

Smart agriculture can benefit from the application of RF, a popular ensemble learning method. It combines many DTs to generate predictions and handles high-dimensional data successfully. Robust predictions, mitigation of overfitting, and handling of high-dimensional data are among the well-known strengths of RF. It is appropriate for challenging classification problems because it can capture non-linear correlations and interactions between features well [57].

6.4.8 Where uncertainty is

$$\sigma = \sqrt{\frac{\sum_{n=1}^{N}(f(x) - f)^2}{N - 1}} \tag{6.4}$$

Finding the best branch for your forest is made easier by the formula, which determines each node's degree of divergence from the estimated real value. In this case, f(x) denotes the DT's output, while f stands for the value of the data point that is being assessed at a particular node.

6.4.9 Clustering and K-nearest neighbors

When utilizing KNN, the method searches the dataset for K close instances. The program will then use its output to evaluate the variable y of the inspection that has to be predicted. To separate two observations, the Euclidean distance formula is:

$$d(x\_i, y\_i) = \surd\left((x\_(i,1) - y\_i,1)\wedge 2 + \cdots + (x\_(i,m) - y\_(i,m))\wedge 2\right) \tag{6.5}$$

For such a distance function, the Euclidean distance is often an acceptable choice if the data are numerical. Our new data point's attribute values are $x\_(i,1)$ through $x\_(i,m)$. The suggested work can calculate the distance $d(x\_i,y\_i)$ between any two data points, y_i and x.

6.4.10 Gaussian Naïve Bayes

One way that Bayes' Theorem stands out is that it considers consecutive events in which prior information affects the starting probability. These possibilities are referred to as "posterior probability" and "prior probability." Prior probability, sometimes referred to as marginal probability, is the first chance of an event occurring in isolation. The posterior probability is the likelihood that something will occur after a data point [58].

The NB approach, which is extensively employed, relies on the mathematical utilization of Bayes' theorem.

$$P(A \mid B) = \frac{P(A \mid B)P(A)}{P(B)} \tag{6.6}$$

The above Bayes theorem states that the dependent feature vectors, x_1 through y_j, and the supplied class variable, y, are related.

$$P(y \mid x_1, \ldots.y_j) = \frac{P(x_1, \ldots.y_j \mid y)P(y)}{P(x_1, \ldots.y_j)} \tag{6.7}$$

Model evaluation is used in ML to select the optimal model. For the ML task, select an evaluation metric such as recall, accuracy, precision, F1 score, or receiver operating characteristic (ROC) area under the curve (AUC). AUC is a crucial metric for evaluating models. Every integer from 0 to 1 satisfies every requirement. AUC near 1 denotes a superior model. Low AUCs indicate poor models, whereas high AUCs suggest strong ones. The AUC for each model's ROC curve was calculated using R's roc tool.

80% of the data in this dataset are training data, while the remaining 20% are testing data. Once the data have been prepared using these techniques, the next step is to find the confusion matrix. The outcomes have established the method's validity. A confusion matrix was used to establish the accuracy.

The algorithms' performance will be compared using equation's CAD prediction accuracy as a benchmark. Using this method will improve the models' accuracy with each iteration, giving each model a total of 10 accuracies at the end of the training phase. We'll compute the accuracy, standard deviation, and average.

$$\text{Accuracy} = \frac{\text{TP} + \text{TN}}{\text{TP} + \text{FP} + \text{FN} + \text{TN}} \tag{6.8}$$

This statistic calculates the proportion of successfully recognized positive data points relative to all positive data points, based on intuition. A higher TPR lowers the number of overlooked good data signals. FPR/Fallout:

$$\text{FRP} = \frac{\text{FP}}{\text{FP} + \text{TN}} \tag{6.9}$$

The primary indicator of classification model performance is AUC-ROC. The capacity of a model to categorize data is measured by AUC-ROC. A model is considered superior if its AUC is larger, in accordance with the grading standards. The correlation and equilibrium between sensitivity and specificity for all possible thresholds in a given test or set of tests are commonly shown visually using AUC-ROC curves.

One technique to gauge a classifier's accuracy is to find the ratio of all of its positive predictions to all of its genuine positives. In summary, precision is the ability to determine what percentage of positive forecasts were realized. You may see the formula below.

$$\text{Precision} = \frac{\text{TP}}{\text{TP} + \text{FP}} \tag{6.10}$$

The recall is calculated by dividing the number of projected positive samples by the total number of samples that were positive.

Recall may be used to determine the proportion of optimistic forecasts that came true.

Below is the formula.

$$\text{Recall} = \frac{\text{TP}}{\text{TP} + \text{FN}} \tag{6.11}$$

Metrics, such as the F-score, take accuracy and memory into account.

$$\text{F1 Score} = \frac{2^{*}\left(\text{Precision}^{*}\text{Recall}\right)}{(\text{Precision} + \text{Recall}} \tag{6.12}$$

6.5 INTEGRATION OF BLOCKCHAIN AND ML APPROACHES FOR THE IESAF

An IESAF that incorporates blockchain and ML techniques offers several benefits-

- Data Integrity and Security
- Decentralized Data Sharing
- Predictive Analytics and Decision Support
- Transactions based on Smart Contracts
- Sustainability and Environmental Monitoring
- Compliance and Certification

Immutable data kept on the blockchain can lessen the amount of paperwork and administrative labor that regulators and farmers must do by offering auditable proof of adherence to certifications and standards. All things considered, the IESAF, when combined with blockchain and ML techniques, provides a strong foundation for enhancing the sustainability, efficiency, and transparency of agricultural systems. IESAF has the potential to transform agricultural data collection, sharing, and use processes, ultimately resulting in more robust and productive agricultural ecosystems. This can be achieved by fusing the security and transparency of blockchain technology with the predictive analytics capabilities of ML.

6.6 PROPOSED ARCHITECTURE OF ML AND BLOCKCHAIN IN THE CONTEXT OF IOT FOR A SMART AGRICULTURAL SYSTEM

Figure 6.1 shows the proposed architecture of blockchain and ML in the context of IESAF.

6.6.1 Data acquisition

A specialized combination of sensors and a processor comprised of hardware circuitry can be used to collect data from agricultural land. Data can be transmitted to distant locations via wireless communication technologies.

6.6.2 Data analysis

There are four methods for analyzing data: preparing the data can improve the accuracy and output of subsequent mining. This process includes data transformation, data integration, and data cleansing. The integrity of the original data must be preserved during the data reduction stage, which treats the reduced data to be more effective while still yielding the same (or nearly the same) analytical findings. To identify patterns in the data, data modeling and discovery employ intelligent algorithms.

It is feasible to use classification, grouping, association, and other analytical approaches. Through data modeling and discovery, intelligent algorithms are applied to the data to identify patterns. Different ML algorithms can be used for data analysis.

6.7 OPPORTUNITIES AND CHALLENGES

Farmers are especially worried about losing personal information because, among other risks, it could be used against them in the commodities market. An insider, such as a disgruntled employee, can leak such knowledge for nefarious reasons or to be sold for a profit. Smart farming data could yield a wealth of private agricultural and economic data on the entire nation. Virtual machines may occasionally be installed in data centers located in other countries due to the global dispersion of cloud data centers. An attacker can disseminate misleading information about a smart farm by saying that crops or cattle are beginning to exhibit illness symptoms. These fake data reports masquerade as real information from the smart farm that is the goal.

6.8 BLOCKCHAIN LAYER

6.8.1 Ethereum

Ethereum, which was founded in 2015 by Vitalik Buterin and a group of fellow entrepreneurs, expands the potential of blockchain technology beyond straightforward bitcoin transactions, enabling intricate programmable contracts and automated procedures. Ether (ETH), the original cryptocurrency of the Ethereum network, is utilized as "gas" to power transactions and calculations in addition to serving as a digital money.

6.8.2 AWS IoT core

It provides the system to develop scalable and effective IoT applications, enabling real-time data transmission between millions of connected devices. With protocols like MQTT, HTTP, and Web Sockets, AWS IoT Core facilitates secure device-to-device communication while maintaining data integrity and privacy [39].

6.8.3 AWS Lambda

Developers can concentrate on building and delivering code more quickly and effectively because AWS Lambda manages the underlying infrastructure and integrates with other AWS services smoothly.

Figure 6.2 represents the flow chart of our system starting from the collection of data from farming land, followed by preprocessing of data, dividing and testing data, data analysis through application of ML, and then sending the relevant data to farmers.

In the context of precision agriculture, where data gathered from farms are utilized to apply ML techniques to deliver farmers actionable insights, this figure appears to represent a data processing process. An in-depth description of every component is given below:

6.8.4 Collecting information from farm land

The first step in the process is gathering data from the farms. This could contain a variety of data, including information on crop growth, temperature, humidity, and soil moisture.

Methods: These data are usually collected via sensors, drones, and other IoT devices, giving precise and up-to-date information about the present condition of the farm.

6.8.5 Data pre-processing:

Data cleaning and organizing from a massive amount of raw data collected from the farming land involves the data preprocessing stage.

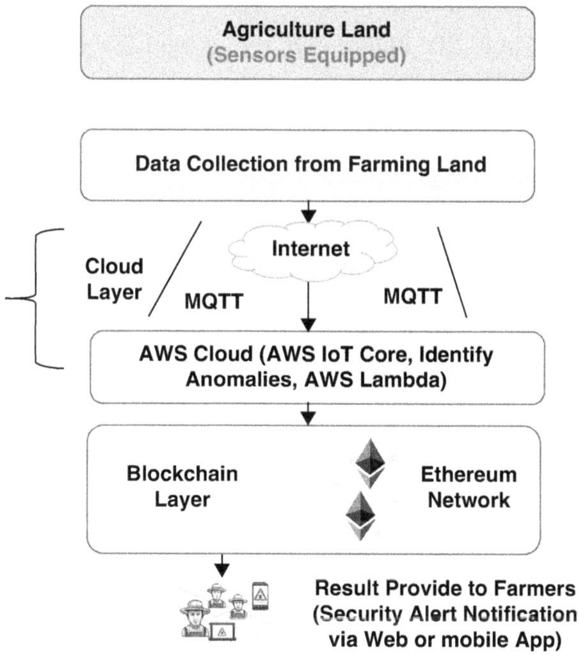

Figure 6.1 Machine learning and blockchain in the context of IoT-based solution in smart farming (proposed architecture).

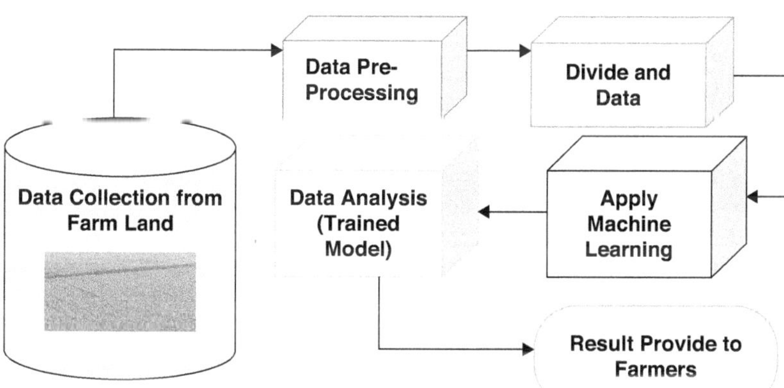

Figure 6.2 System flowchart.

6.8.6 Divide and test data

After data pre-processing, it is divided into training and testing datasets.

6.8.7 Apply the algorithm for ML

In this stage, various ML algorithms are applied for the optimization of agricultural parameters.

6.8.8 Data analysis (trained model)

Predicting results like agricultural production, optimum irrigation schedules, or potential crop issues is made easier with the use of analysis.

6.8.9 Farmers received the outcome

Through the user interface, a farmer can get information about the agriculture land.

6.9 INTEGRATION CONCERNS FOR SMART AGRICULTURE

Integrating smart agriculture technologies into traditional farming practices can offer numerous benefits, but it also comes with several concerns and challenges that need to be addressed. Below are some key integration concerns for smart agriculture:

- Security and Data Management
- Interoperability
- Privacy and Security
- Scalability

A comprehensive strategy involving cooperation between farmers, technology suppliers, legislators, and researchers is needed to address these issues. In addition to making smart agriculture technologies more dependable, accessible, and user-friendly, efforts should be directed to ensuring that they favorably impact social justice and sustainability in the agricultural industry.

6.10 FUTURE RESEARCH DIRECTIONS RELATED TO SMART AGRICULTURE

The advancement of smart agriculture and the resolution of new issues depend heavily on the future paths of research. The following are a few interesting topics for further study:

Technologies for Precision Agriculture: Create cutting-edge sensing tools, such as drones, LiDAR, and hyperspectral imaging, to precisely and instantly monitor environmental factors, soil quality, and crop health. Combine these technologies with ML and data analytics to maximize agricultural yield and resource management.

Sensor Networks and the IoT: Examine the set of sensors used in the IoT-based system.

Blockchain and Distributed Ledger Technologies: Examine how blockchain and distributed ledger technology could improve agricultural supply networks' traceability, transparency, and trustworthiness. Investigate uses for smart contracts to automate agreements and transactions among stakeholders, as well as food provenance, certification, and compliance tracking.

Advanced ML and AI for Precision Farming: Advance techniques in data analytics and AI are utilized to analyze vast amounts of agricultural data. Create predictive models using multi-modal data sources and heterogeneous data integration for agricultural production forecasts, disease detection, pest control, and climate resilience evaluation.

Digital Agricultural Extension Services: Create decision support tools and digital agricultural extension services to offer farmers individualized advice and recommendations. Provide actionable insights, training materials, and best practices catered to regional settings and farmer preferences by integrating mobile applications, web platforms, and AI-powered chatbots.

Socioeconomic Research and Policy: Undertake multidisciplinary investigations concerning the economic, policy, and socio-cultural facets of the adoption and dissemination of smart agriculture. Examine how technical advancements affect rural lives, economic inequality, and food security. Then, devise plans of action to advance inclusive and equitable agricultural development.

Resilience and Disaster Management: Examine methods for enhancing resilience and frameworks for disaster management to lessen the effects of extreme weather events, climatic variability, and natural catastrophes on agricultural output. Provide risk assessment instruments, early warning systems, and adaptive tactics to strengthen resilience and guarantee food security in areas that are susceptible.

Knowledge Sharing and Collaborative Platforms: Encourage cooperation and information exchange between farmers, researchers, policymakers, and business partners, as well as other stakeholders in the agricultural ecosystem. Create innovation centers, open data projects, and cooperative platforms that support co-creation, experimentation, and technology transfer in smart agriculture.

6.11 CASE STUDY ON SMART AGRICULTURE

Case Study: Blockchain and ML in Smart Agriculture in the Context of the IoT

In this case study, we will investigate how Geo Farms, a smart agricultural firm, uses blockchain, the IoT, and ML to maximize supply chain transparency, guarantee food safety, and improve crop management.

A sizable agricultural company that specializes in organic vegetables is called Geo Farms. The organization strives to optimize agricultural yields while minimizing resource consumption and environmental impact, all the while maintaining a commitment to sustainability and innovation. Geo Farms uses cutting-edge technologies like blockchain, ML, and IoT to accomplish these objectives.

The characteristics of this system are as follows:

IoT-enabled Sensor Network: Geo Farms uses IoT sensors to track temperature, humidity, and other environmental variables in real time throughout its fields. Data from these sensors is sent to a centralized cloud platform for processing and analysis.

ML **for Crop Management:** To improve crop management techniques, ML algorithms examine data gathered from IoT sensors. Based on historical and current data, ML algorithms forecast insect outbreaks, identify nutrient deficits, and estimate crop water requirements.

Blockchain-enabled Traceability: The blockchain records every step of the production process, from planting to harvesting to distribution, giving stakeholders and customers access to transparent and verifiable information.

Smart Contracts for Agreements: Using blockchain technology, smart contracts automate and uphold contracts between Geo Farms and its distributors, retailers, and suppliers.

Food Safety Assurance: In the case of a food safety problem or recall, Geo Farms can promptly track the origin of the product, thanks to blockchain-based traceability. Customers' trust and confidence in the safety and integrity of the company's goods are increased by accessible access to manufacturing data and quality certificates.

Increased Crop Yields: Geo Farms optimizes resource management techniques to obtain increased crop yields and lower input costs by utilizing ML algorithms and Blockchain.

6.12 CONCLUSION

Blockchain technology and ML are complementary to one another, providing a strong foundation for innovation and industry change. Within the framework of the IoT, this integration helps businesses to fully utilize distributed ledger technology, data analytics, and connected devices to build systems that are more intelligent, effective, and transparent. In the future, it is anticipated that more study and advancement in the IoT ecosystem's integration of blockchain technology and ML will open even more opportunities, allowing businesses to solve difficult problems, generate fresh business ideas, and provide value to society. Organizations that embrace IoT, ML, and blockchain and investigate creative use cases stand to benefit from a competitive edge and beneficial influence in the digital age as these technologies become more widely used.

REFERENCES

[1] "State of Agriculture in India" (2025) [Online], https://prsindia.org/policy/analytical-reports/state-agriculture-india

[2] "Farmer's Welfare in India" (2025) [Online], nabard.org/auth/writereaddata/tender/1710224557farmers-welfare-in-india-a-state-wise-analysis.pdf

[3] L. Ahmad and F. Nabi, *Agriculture 5.0: Artificial intelligence, IoT, and Machine Learning*. 2021. doi: 10.1201/9781003125433.

[4] K. N. Dhawale, "Review on IoT based smart agriculture system," *Int. J. Res. Appl. Sci. Eng. Technol.*, vol. 11, no. 3, pp. 1–5, 2019.

[5] Amandeep, A. Bhattacharjee, P. Das, D. Basu, S. Roy, S. Ghosh, S. Saha, S. Pain, S. Dey, and T. K. Rana, "Smart farming using IOT," in *Proc. 2017 8th IEEE Annual Inf. Technol., Electron. Mobile Commun. Conf. (IEMCON)*, 2017, pp. 278–280.

[6] L. Zheng, M. Li, C. Wu, H. Ye, R. Ji, X. Deng, Y. Che, C. Fu, and W. Guo, "Development of a smart mobile farming service system," *Math. Comput. Model.*, vol. 54, pp. 1194–1203, 2011.

[7] "Smart agriculture management system using internet of things," *TELKOMNIKA (Telecommunication Computing Electronics and Control)*, 2020.

[8] C. C. Baseca, S. Sendra, J. Lloret, and L. García, "An IoT service-oriented system for agriculture monitoring," in *Proc. 2017 IEEE Int. Conf. Commun. (ICC)*, 2017, pp. 1–6.

[9] S. Qazi, B. A. Khawaja, and Q. U. Farooq, "IoT-equipped and AI-enabled next generation smart agriculture: A critical review, current challenges and future trends," *IEEE Access*, vol. 10, pp. 21219–21235, 2022.

[10] V. Nandal and S. Dahiya, "IoT based energy-efficient data aggregation wireless sensor network in agriculture: A review," *Psychol. Educ.*, vol. 58, no. 1, pp. 2985–3007, 2021.

[11] I. Mat, M. R. Mohd Kassim, A. N. Harun, and I. M. Yusoff, "Smart agriculture using the internet of things," in *Proc. 2018 IEEE Conf. Open Syst. (ICOS)*, 2018, pp. 54–59.

[12] M. M. Subashini, S. Das, Heble, U. Raj, and R. Karthik, "Internet of things based wireless plant sensor for smart farming," *Indones. J. Electr. Eng. Comput. Sci.*, vol. 10, pp. 456–468, 2018.

[13] V. K. Quy, N. V. Hau, D. V. Anh, N. M. Quy, N. T. Ban, S. Lanza, G. Randazzo, and A. Muzirafuti, "IoT-enabled smart agriculture: Architecture, applications, and challenges," *Appl. Sci.*, vol. 12, no. 7, p. 3396, 2022. Doi: 10.3390/app12073396
[14] A. Dahane, R. Benameur, K. Bouabdellah, and A. E. Benyamina, "An IoT based smart farming system using machine learning," in *Proc. 2020 Int. Symp. Netw., Comput. Commun. (ISNCC)*, 2020, pp. 1–6.
[15] N. Ahmed, D. De, and I. Hussain, "Internet of things (IoT) for smart precision agriculture and farming in rural areas," *IEEE Internet Things J.*, vol. 5, pp. 4890–4899, 2018.
[16] K. N. Bhanu, S. Mahadevaswamy, and H. J. Jasmine, "IoT based Smart system for enhanced irrigation in agriculture," in *Proc. 2020 Int. Conf. Electron. Sustain. Commun. Syst. (ICESC)*, 2020, pp. 760–765.
[17] N. K. Jadav, T. Rathod, R. Gupta, S. Tanwar, N. Kumar, and A. Alkhayyat, "Blockchain and artificial intelligence-empowered smart agriculture framework for maximizing human life expectancy," *Comput. Electr. Eng.*, vol. 105, p. 108486, 2023.
[18] A. Aliyu and J. Liu, "Blockchain-based smart farm security framework for the internet of things," *Sensors*, vol. 23, no. 18, p. 7992, 2023.
[19] S. Vyas, M. Shabaz, P. Pandit, L. R. Parvathy, and I. Ofori, "Integration of artificial intelligence and blockchain technology in healthcare and agriculture," *J. Food Qual.*, 4228448, 11 pp., 2022. doi: 10.1155/2022/4228448
[20] R. Yadav, R. Saroj, A. Kumar Verma, and A. Kumar Mishra, "A survey of IoT and machine learning based monitoring of the growth of crops using blockchain technology," in *Proc. 2023 Int. Conf. IoT, Commun. Autom. Technol. (ICICAT)*, 2023, pp. 1–7.
[21] M. Maduranga and R. Abeysekera, "Machine learning applications in IoT based agriculture and smart farming: A review," *Int. J. Eng. Appl. Sci. Technol.*, vol. 04, no. 12, pp. 24–27, 2020. DOI: 10.33564/IJEAST.2020.v04i12.004
[22] J. Doshi, T. Patel, and S. K. Bharti, "Smart farming using IoT, a solution for optimally monitoring farming conditions," in *Proc. EUSPN/ICTH*, 2019.
[23] S. H. Awan, "Role of internet of things (IoT) with blockchain technology for the development of smart farming," *J. Mech. Contin. Math. Sci.*, 14(5), 2019. Doi: 0.26782/jmcms.2019.10.00014
[24] A. Ramesh Kumar, K. B. Archana, and P. Medhinya, "Machine learning for IoT-based smart farming," *J. Adv. Zool.*, vol. 44, no. S-3, pp. 1294–1298, Nov. 2023. doi: 10.17762/jaz.v44is-3.1494.
[25] R. Chaganti, V. Varadarajan, V. S. Gorantla, T. R. Gadekallu, and V. Ravi, "Blockchain-based cloud-enabled security monitoring using Internet of Things in smart agriculture," *Future Internet*, vol. 14, p. 250, 2022.
[26] M. Gayatri, J. Jayasakthi, and G. Mala, "Providing smart agricultural solutions to farmers for better yielding using IoT," in *Proc. 2015 IEEE Technol. Innov. ICT Agric. Rural Dev. (TIAR)*, 2015, pp. 40–43.
[27] A. Kaushal, "Overview of IoT and blockchain technology in smart agriculture," *Int. J. Res. Appl. Sci. Eng. Technol.*, vol. 9, no. VII, pp. 1908–1913, 2021. Doi: 10.22214/ijraset.2021.36765
[28] P. Singh and N. Singh, "Blockchain with IoT and AI: A review of agriculture and healthcare," *Int. J. Appl. Evol. Comput.*, vol. 11, pp. 13–27, 2020.

[29] M. Torky and A. E. Hassanein, "Integrating blockchain and the internet of things in precision agriculture: Analysis, opportunities, and challenges," *Comput. Electron. Agric.*, vol. 178, p. 105476, 2020.
[30] T. H. Pranto, A. A. Noman, A. Mahmud, and A. B. Haque, "Blockchain and smart contract for IoT enabled smart agriculture," *PeerJ Comput. Sci.*, vol. 7, p. e407, Mar. 2021. Doi: 10.7717/peerj-cs.407.[31]
[31] S. D. Padiya, A. Raut, B. Mittal, M. Patel, and V. Polshettiwar, "Machine learning and IoT applications in agriculture," *Int. J. Adv. Res. Sci. Commun. Technol.*, vol. 3, no. 6, pp. 287–293, 2023. Doi: 10.48175/IJARSCT-9416
[32] L. Gupta, S. Malhotra, and A. Kumar, "Study of applications of internet of things and machine learning for smart agriculture," in *Proc. 2022 IEEE Int. Conf. Curr. Dev. Eng. Technol. (CCET)*, 2022, pp. 1–5.
[33] R. A. Ahmed, E. E. Hemdan, W. El-shafai, Z. A. Ahmed, E. M. El-Rabaie, and F. E. Abd El-Samie, "Climate-smart agriculture using intelligent techniques, blockchain and Internet of Things: Concepts, challenges, and opportunities," *Trans. Emerg. Telecommun. Technol.*, vol. 33, no. 11, e4607, 2022. doi:10.1002/ett.4607
[34] M. W. P. Maduranga and R. Abeysekera, "Machine learning applications in IoT based agriculture and smart farming: A review," *Int. J. Eng. Appl. Sci. Technol.*, vol. 04, no. 12, pp. 24–27, May 2020. doi: 10.33564/ijeast.2020.v04i12.004.
[35] A. Biswas, T. Akhund, M. J. Ferdous, S. Kar, A. Anis, and S. A. Shanto, "BIoT: Blockchain based smart agriculture with Internet of Things," in *Proc. 2021 5th World Conf. Smart Trends Syst. Secur. Sustain.* (WorldS4), 2021, pp. 75–80.
[36] O. Friha, M. A. Ferrag, L. Shu, L. Maglaras, and X. Wang, "Internet of Things for the future of smart agriculture: A comprehensive survey of emerging technologies," *IEEE/CAA J. Autom. Sinica*, vol. 8, pp. 718–752, 2021.
[37] D. Kumar and R. K. Dwivedi, "Blockchain and IoT based smart agriculture and food supply chain system," in *Proc. 2023 Int. Conf. Intell. Innov. Technol. Comput. Electr. Electron. (IITCEE)*, 2023, pp. 755–761.
[38] S. Bhatia, Z. A. Jaffery, and S. Mehfuz, "A comparative study of wireless communication protocols for use in smart farming framework development," in *Proc. 2023 3rd Int. Conf. Intell. Commun. Comput. Tech. (ICCT)*, 2023, pp. 1–7.
[39] S. Bhatia, B. Mallikarjuna, D. Gautam, U. Gupta, S. Kumar, and S. Verma, "The future IoT: The current generation 5G and next generation 6G and 7G technologies," in *Proc. 2023 Int. Conf. Device Intell. Comput. Commun. Technol. (DICCT)*, 2023, pp. 212–217.
[40] S. Bhatia, Z. A. Jaffery, S. Mehfuz, and N. Goel, "Integration of WSN and IoT: Wireless networks architecture and protocols–A way to smart agriculture," in *Handbook of Research on Machine Learning-Enabled IoT for Smart Applications Across Industries*, IGI Global, 2023, pp. 435–455.
[41] K. Haseeb, I. Ud Din, A. Almogren, and N. Islam, "An energy efficient and secure IoT-based WSN framework: An application to smart agriculture," *Sensors (Basel)*, vol. 20, no. 7, p. 2081, Apr. 2020, doi: 10.3390/s20072081.
[42] P. T. Ganai, A. Bag, A. Sable, K. H. Abdullah, S. Bhatia, and B. Pant, "A Detailed Investigation of implementation of internet of things (IoT) in cyber security in healthcare sector," in *Proc. 2022 2nd Int. Conf. Adv. Comput. Innov. Technol. Eng. (ICACITE)*, 2022, pp. 1571–1575.

[43] A. Msd, J. Kuppili, and N. A. Manga, "Smart farming system using IoT for efficient crop growth," in *Proc. 2020 IEEE Int. Students' Conf. Electr., Electron. Comput. Sci. (SCEECS)*, 2020, pp. 1–4.
[44] O. Elijah, T. A. Rahman, I. Orikumhi, C. Y. Leow, and M. H. Hindia, "An overview of Internet of Things (IoT) and data analytics in agriculture: Benefits and challenges," *IEEE Internet Things J.*, vol. 5, pp. 3758–3773, 2018.
[45] C. Yoon, M. Huh, S. Kang, J. Park, and C. Lee, "Implement smart farm with IoT technology," in *Proc. 2018 20th Int. Conf. Adv. Commun. Technol. (ICACT)*, 2018, pp. 749–752.
[46] S. Bhatia, N. Goel, V. Ahlawat, B. B. Naib, and K. Singh, "A comprehensive review of IoT reliability and its measures: perspective analysis," in *Handbook of Research on Machine Learning-Enabled IoT for Smart Applications Across Industries*, IGI Global, 2023, pp. 365–384.
[47] S. K. Omar, "Internet of Things (IoT) for smart farming: A systematic review," *Int. J. Comput. Appl.*, vol. 174, no. 27, pp. 47–54, 2021. Doi: 10.5120/ijca2021921182
[48] S. Bhatia, Z. A. Jaffery, and S. Mehfuz, "Development and analysis of IoT-based smart agriculture system for heterogeneous nodes," in *Proc. 2023 Int. Conf. Recent Adv. Electr., Electron. Digit. Health Technol. (REEDCON)*, 2023, pp. 62–67.
[49] P. Jaiswal, N. Rajput, R. Kumar, S. Bhatia, and R. Kumar, "Drone automation with smart irrigation," in *Futuristic Sustainable Energy & Technology: Proc. Int. Conf. Futur. Sustain. Energy & Technol. (ICFSE)*, CRC Press, 2022, p. 467.
[50] R. Umrao, R. Srivastava, S. Shama, D. Kumar, and S. Bhatia, "Precision farming system using IoT," in *Futuristic Sustainable Energy & Technology: Proc. Int. Conf. Futur. Sustain. Energy & Technol. (ICFSE)*, CRC Press, 2022, p. 477.
[51] H. Y. Chen, K. Sharma, C. Sharma, and S. Sharma, "Integrating explainable artificial intelligence and blockchain to smart agriculture: research prospects for decision making and improved security," *Smart Agric. Technol.*, vol. 6, p. 100350, 2023.
[52] G. K. Akella, S. Wibowo, S. Grandhi, and S. Mubarak, "A systematic review of blockchain technology adoption barriers and enablers for smart and sustainable agriculture," *Big Data Cogn. Comput.*, vol. 7, no. 2, p. 86, 2023.
[53] R. Jothikumar, "Applying blockchain in agriculture: A study on blockchain technology, benefits, and challenges," in *Deep Learning and Edge Computing Solutions for High Performance Computing*, Springer, Cham, 2021, pp. 167–181.
[54] N. M. Chacko, V. G. Narendra, M. Balachandra, and S. Rathinam, "Exploring IoT-blockchain integration in agriculture: An experimental study," *IEEE Access*, vol. 11, pp. 130439–130450, 2023.
[55] G. Chiaraluce, D. Bentivoglio, A. Finco, M. Fiore, F. Contò, and A. Galati, "Exploring the role of blockchain technology in modern high-value food supply chains: Global trends and future research directions," *Agric. Food Econ.*, vol. 12, no. 1, p. 6, 2024.
[56] A. Pakseresht, S. A. Kaliji, and K. Hakelius, "Blockchain technology characteristics essential for the agri-food sector: A systematic review," *Food Control*, vol. 2024, p. 110661, 2024.

Chapter 7

Empowering Indian agriculture through intelligent automation

A fuzzy wireless sensor network approach

Jayashree S. Awati, Vijay H. Kalmani, Mahesh Kumbar, and Nagaraj V. Dharwadkar

7.1 INTRODUCTION

As the backbone of the Indian economy and the means of livelihood for many, agriculture is still affected to some extent by its drawbacks, even after improvements in technology, which takes place in open fields, such as declining productivity, erratic weather conditions, improper use of chemicals, and insufficient modernization. Protected farming practices, such as greenhouse cultivation, have received a fair share of technological innovations. However, traditional farming continues to rely on age-old techniques that impart inefficiencies and environmental risks to its operations. To make matters worse, the emigration of the younger generation from agriculture creates a further distance from the sector's longevity.

Thus, the use of Fuzzy Wireless Sensor Networks (FWSNs) can help bridge the technological divide in Indian agriculture. Real-time data acquisition, combined with smart decision-making and automation, will help FWSNs maximize resource use, optimize irrigation efficiency, and minimize the destruction of crops by pests. On the one hand, farming efficiency is enhanced, and on the other hand, it has favorable socio-economic impacts. Agricultural productivity empowers farmers' income security and mitigates their dependence on fluctuating market forces. By maintaining such fields, they mostly enhance soil quality through less over-irrigation or the use of pesticides as a means of environmental preservation in precision farms with FWSNs.

Socially and economically, the inclusion of FWSNs in agriculture has created rural employment opportunities for technologically innovative farm workers trained in sensor technology, data analytics, and automated farm solutions. This can also motivate youth involvement in agriculture by making farming more technologically attractive. Governmental policy programs favoring precision agriculture, together with cheap automation offerings, might jointly provide an additional impetus toward large-scale adoption, thereby securing India's food and providing for the rural economy.

DOI: 10.1201/9781003494782-7

This study analyzes the transformation of Indian agriculture with the application of FWSNs to overcome heavy farming constraints through automation and substantiates its effect through statistical verification. With the aim of equipping farmers with smart and low-cost solutions, this research intends to modernize and render sustainable Indian agriculture for long-term economic and environmental benefits.

7.2 METHODOLOGY

An innovative methodology addresses the entire challenge faced by an Indian farmer by enabling intelligent automation through fuzzy sensor networks. The following steps were performed:

i. **Identification of the Research Problem**
   Step one involves the identification of pertinent issues in agriculture. These are the unique challenges that farmers face, concerns about the environment of a given area, and the need for modernization.
ii. **Survey research and analysis of the problem**
   The survey method was used to understand farmers' problems. A structured questionnaire was prepared and administered to farmers in Karandwadi Village, Taluka, Walwa District, Sangli, Maharashtra, India. The survey provided detailed insights through an evaluation of several agricultural problems faced by farmers.
iii. **Literature Review**
   This study presents a literature analysis to understand the existing research and solutions related to farming challenges, wireless technologies, and automation.
iv. **Designing Solutions**
v. According to surveys and literature review findings, solutions are devised that focus on the identified problems. Solutions are targeted at a specific case and the use of wireless technology for intelligent automation.
vi. **Experimental Research**
   This research consists of genuine experiments with automation utilizing WSNs in the design and implementation process. Such experiments were conducted in the field of Karandwadi Village.
vii. **Data Collection and Analysis**
   The experimental factors in data collection were the machine operation of the automatic insect killer, identification of the harmful insects of the crop, and maintenance of the correct temperature in the agricultural area. Statistical methods used were analysis of variance (ANOVA), receiver operating characteristic (ROC) analysis, and correlation analysis.

7.3 OBJECTIVES OF THE WORK

i. **Assessing Agricultural Issues:** To evaluate and comprehend the agricultural problems faced by farmers in India. Such problems include the identification of specific problems in the study area.
ii. **Hypothesis Testing:** Testing the hypotheses formed during research, particularly the effectiveness of intelligent automation using sensor networks in solving farm-related problems.
iii. **Automation Development:** Design, development, and implementation of automation involving WSNs to support performance improvement in different areas of agricultural practices.
iv. **Routing Protocol Design:** To improve the performance and utilization of routing protocols for WSNs, specifically for applications in agricultural research and automation.
v. **Experimental Data Analysis:** Analysis of results from experiments evaluating the use of automation solutions is through the use of statistical methods that are used not only to estimate the impact of improved farming practices but also to achieve a competitive advantage.

7.4 LITERATURE REVIEW

The spurt of advancement in WSN, Internet of Things (IoT), and artificial intelligence (AI)-based automation has impacted precision farming. Several investigations have been conducted on robotics, energy-efficient protocols, decision-making algorithms, and real-time monitoring systems, with the aim of increasing agricultural productivity. The table highlights the important research contributions in this area.

Table 7.1 provides an overview of studies conducted to optimize agriculture using automation, IoT, and sensor-based technologies. The findings of these studies can be used to create more efficient, sustainable, and smart farming systems.

7.5 PROBLEM IDENTIFICATION, DATA COLLECTION, AND STATISTICAL TESTING

Modern automation exists in the agricultural sector, yet persistent challenges plague farmers. To address these issues, a meticulous examination of farmers' problems was conducted using the case study approach. Farmer interviews and questionnaires were used to gather essential data. A cost-effective automation system was designed to mitigate these problems, involving the development of a WSN, protocol implementation, and decision support system. Statistical tools were harnessed for problem analysis,

Table 7.1 Summary of research contributions in agricultural IoT and wireless sensor networks (WSN)

Author(s) and Year	*Research focus*	*Methodology*	*Key findings*	*Source*
Bechar et al. [1]	Agricultural robotics and fuzzy logic	Designed four-wheeled farm robots and a forest-navigating robot	Improved decision-making in unstructured farm environments	Elsevier
Botta et al. [2]	Cloud IoT in agriculture	Surveyed cloud computing and IoT integration	Defined cloud IoT paradigm and discussed open-source platforms	IEEE
Sanchez et al. [3]	Precision agriculture and WSN	Implemented video surveillance for crop monitoring	Enabled real-time farm communication and security	Springer
Kharazian et al. [4]	Energy-efficient WSN routing	Tested Leach-C protocol in simulations	Outperformed the conventional Leach algorithm in energy efficiency	IEEE
Delava et al. [5]	Cluster-based routing in WSN	Proposed SLGC algorithm for grid-based clustering	Reduced energy consumption in sensor networks	Elsevier
Sivakumar et al. [6]	Low-power WSN in agriculture	Developed a stable WSN system for flood prevention	Demonstrated low-power technology effectiveness	Springer
Rad et al. [7]	Cyber-physical systems (CPS)	Explored CPS architecture with Wi-Fi, Zigbee, and RF	Provided a robust framework for smart farming	IEEE
Deepika and Rajapirian [8]	Environmental monitoring using WSN	Designed a web-based data management system	Enabled secure sensor data storage and visualization	Elsevier
Ye et al. [9]	WSN deployment for farming	Implemented a 1km x 1km WSN network with IRIS hardware	Showcased real-time data acquisition and processing	Springer
Romanov et al. [10]	Real-time precision farming	Designed an IEEE 802.15.4-based WSN monitoring system	Enhanced agricultural environment analysis	IEEE
Dasgupta and Darshan [11]	Smart agriculture WSN	Simulated LabVIEW-based sensor network	Enabled real-time crop monitoring and data transmission	Elsevier

and the research was conducted in Karandwadi village, Maharashtra, India, with a focus on crops, such as sugarcane and vegetables. A total of 53 farmers participated in the survey, which was conducted before, during, and after the implementation of automation. Three crucial variables related to automatic insect killers and crop-harming insects were selected for in-depth statistical analysis: ANOVA, ROC, and correlation testing.

7.5.1 Problem identified

This study identified critical challenges in farming activities, mainly the lack of measurement and analysis of the effectiveness of agricultural practices by farmers. Several factors have been recognized to affect the success of automation in agriculture.

i. **Farmer Knowledge:** This encompasses aspects such as agricultural expertise, experience, motivation, income, and practices.
ii. **Automation Variables:** These variables include the usefulness, cost, and expertise of automation systems.
iii. **Societal Variables:** Factors such as farmer practices, group decision-making, and crop variety
iv. **Environmental Variable:** These changes in the environment and their impact on crops.

7.5.2 Data collection

Primary data collection methods were employed, using questionnaires designed to assess farmers' problems. The questionnaire included 25 questions covering various aspects, including the identification of crop-harming insects, use of insect-killing automation, and maintenance of agricultural temperatures. Hypotheses were formulated based on the collected data, with H1 focusing on the harmfulness of insects at specific times, and H2 examining the impact of automation on crop yield and cost reduction.

7.5.3 Statistical analysis

Statistical tools are crucial for evaluating formulated hypotheses.

H1: At a particular time, insects are harmful to crops.
H2: Automation to kill harmful insects helps farmers to yield more at less cost.

7.5.4 ANOVA testing

One-way repeated measures ANOVA was applied to assess variations between questionnaire results obtained at different stages of the research: before automation, during automation implementation, and after crop yield.

This analysis aimed to gauge the impact of harmful insects and temperature maintenance on the crops. SPSS software, a statistical tool, facilitated this assessment.

Table 7.2 displays the ANOVA test results for harmful insect killers, specifically for insect harmfulness at particular times. The p-values for all variables were significantly greater than 0.05, indicating no significant difference between the groups. Consequently, the null hypothesis was accepted, confirming that insects are harmful at specific times.

Table 7.3 presents the ANOVA test results for the effect of automation in killing harmful insects on crop yield and cost reduction. As in the previous table, the *p*-values exceeded 0.05, implying no substantial variation between the groups. The null hypothesis was again accepted, validating the hypothesis that automation aids in increasing yields while reducing costs.

Table 7.2 ANOVA testing for harmful insect killer

Test result variable(s)		*Sum of squares*	*df*	*Mean square*	*F*	*Sig.*
Insects at a particular time are harmful to crops	Between groups	0.354	2	0.175	0.721	0.492
	Within groups	12.458	51	0.245		
	Total	12.814	53			
Recommend a harmful insect killer	Between groups	0.365	2	0.183	2.607	0.085
	Within groups	3.563	51	0.071		
	Total	3.927	53			
Implemented a harmful insect killer	Between groups	0.498	2	0.251	2.337	0.106
	Within groups	5.431	51	0.107		
	Total	5.928	53			

Table 7.3 ANOVA testing for automation in killing harmful insects helps farmers obtain more yield with less cost

Test result variable(s)		*Sum of squares*	*df*	*Mean square*	*F*	*Sig.*
Insect killers helpful to farmers	Between groups	0.058	2	0.029	0.148	0.861
	Within groups	10.033	51	0.195		
	Total	10.092	53			
Recommend a harmful insect killer for a higher yield	Between groups	0.003	2	0.001	0.071	0.932
	Within groups	0.078	51	0.018		
	Total	0.981	53			
Implemented a harmful insect killer with less cost.	Between groups	0.178	2	0.887	0.355	0.701
	Within groups	12.637	51	0.247		
	Total	12.816	53			

7.5.5 Correlation testing

Correlation testing was performed to measure the relationships between the variables.

Table 7.4 displays the correlation between the variables related to harmful insects and the recommendation of insect killers for better yields and cost savings. The majority of the correlations were positive, indicating that an increase in one variable corresponded to an increase in the other. This finding supports the hypothesis that insect killers enhance yield and reduce costs.

7.5.6 ROC curve testing

ROC testing was used to explore the interplay between variables, particularly harmful insect killers, and their impact on yields and costs.

The ROC curve shown in Figure 7.1 demonstrates a balanced relationship between variables, showing the positive effect of insect killers with a proper cutoff time. This analysis revealed that automation in agriculture has a statistically proven positive impact on cash crops, enhancing yield while reducing costs.

7.6 BLOCK DIAGRAM OF THE RESEARCH WORK

The system utilizes WSNs, Fuzzy Logic, and Automation to improve farm productivity, reduce resource wastage, and enable precision farming. The block diagram is graphically illustrated in Figure 7.2, which shows

Table 7.4 Correlation for a harmful insect killer

Test result variable(s)		*Insects at a particular time are harmful to crops*	*Recommend a harmful insect killer for a higher yield*	*Implemented a harmful insect killer with less cost*
Insects at a particular time are harmful to crops	Pearson correlation	1	0.098	0.033
	Sig. (2-tailed)		0.476	0.806
	N	53	53	53
Recommend a harmful insect killer for a higher yield	Pearson correlation	0.099	1	0.191
	Sig. (2-tailed)	0.476		0.163
	N	53	53	53
Implemented a harmful insect killer with less cost.	Pearson correlation	0.035	0.0191	1
	Sig. (2-tailed)	0.806	0.0165	
	N	53	53	53

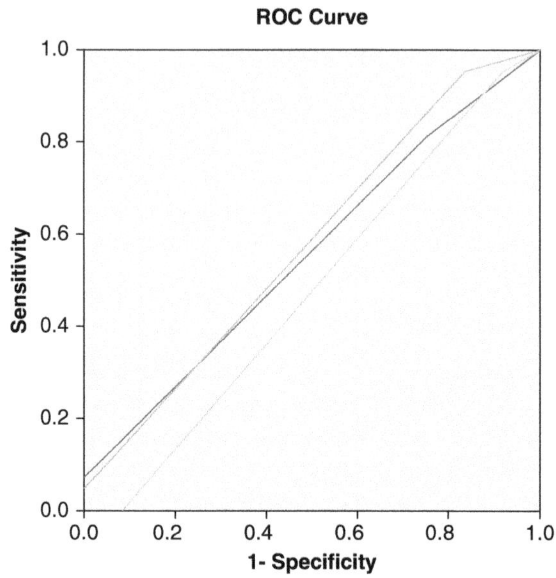

Figure 7.1 ROC curve.

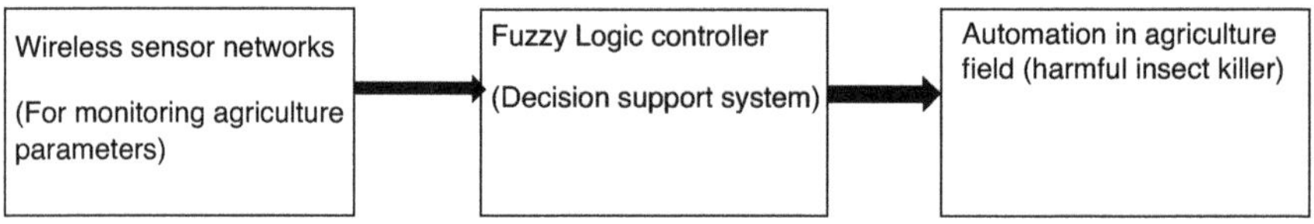

Figure 7.2 Block diagram of the research work.

the interface of these technologies in an integrated manner for intelligent, data-based decision-making for sustainable farming.

1. **WSNs—Real-Time Agricultural Monitoring**

   These WSNs are used to form a backbone for the real-time collection of sensing data from the environment: crop health. Different sensors can be installed strategically throughout the field to measure the following:

   i. **Soil Moisture:** Determines water content to maximize irrigation.
   ii. **Temperature and Humidity:** Aids in controlling the microclimate conditions.
   iii. **Light Intensity:** Aids in measuring plant photosynthesis efficiency.
   iv. **Pest Detection:** Detects early pest activity to avoid infestations.

These sensors send data to a central processing unit continuously through a wireless network, avoiding the intricacies of wired configurations and facilitating energy-saving and seamless communication between farm locations.

2. **Fuzzy Logic—Intelligent Decision-Making for Precision Farming**
   Processed sensor data are processed by a Fuzzy Logic Controller (FLC), which improves decision making by managing uncertainties and fluctuations in environmental conditions. Fuzzy logic, unlike traditional rule-based decision making, does the following:
   i. Optimizes **Irrigation**, rather than applying a fixed moisture level, temperature, humidity, and past irrigation history to make the appropriate watering schedule.
   ii. Improves **Pest Control:** Examine pest detection patterns and environmental factors to initiate timely action, minimizing pesticide misuse.
   iii. Improves **Resource Efficiency:** Provides adaptive action based on varying field conditions, resulting in zero water, fertilizer, or pesticide waste.
3. **Automation—Adoption of Smart Farming Technologies**
   The last phase is automation, wherein the actuators and control systems perform actions based on fuzzy-logic-inspired knowledge. Some of the most important automation processes are as follows:
   i. **Automated Irrigation System:** Smart valves control water supply, avoiding over-irrigation and drought stress.
   ii. **Pest Control System:** Smart pesticide dispensers and organic repellents are turned on at a particular time, minimizing chemical applications.

7.7 DESIGNED AND DEVELOPED WSN

A WSN segment was meticulously designed and developed using microcontrollers and ZigBee modules. As depicted in Figure 7.3, this segment incorporates three Personal Area Networks (PAN) and nine wireless sensor nodes. To ensure efficient functionality, a custom routing algorithm was formulated, developed, and rigorously tested in an agricultural setting. The routing algorithm devised is a fusion of the Bellman–Ford routing algorithm and the leech algorithm. These WSNs monitor vital agricultural parameters, such as temperature, humidity, wind direction, and rainfall, relaying these data to the FLC for informed farming decisions.

The base station, as shown in Figure 7.4, is equipped with ZigBee modules and interfaces with a PC or laptop for data collection and display via a graphical user interface (GUI) built using QT.

QT Creator is used to design the GUI, making it user-friendly and efficient for data monitoring, as shown in Figure 7.5.

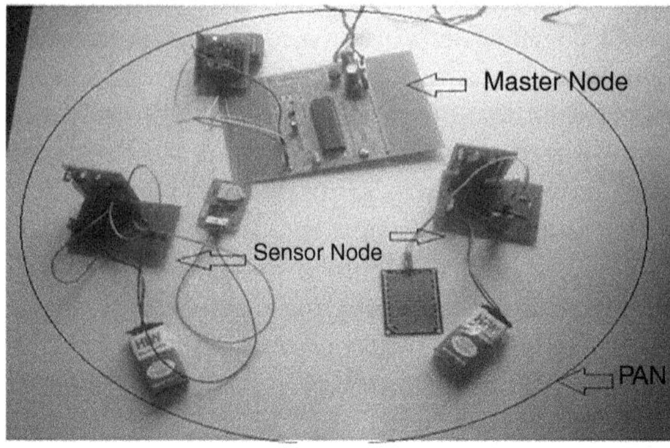

Figure 7.3 Developed wireless sensor network hardware of one PAN.

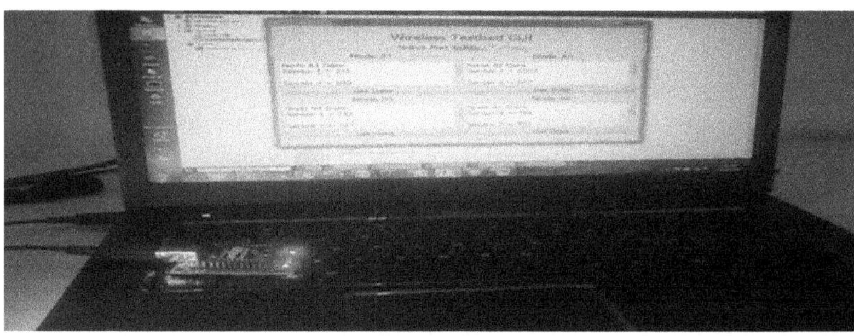

Figure 7.4 Hardware of the base station.

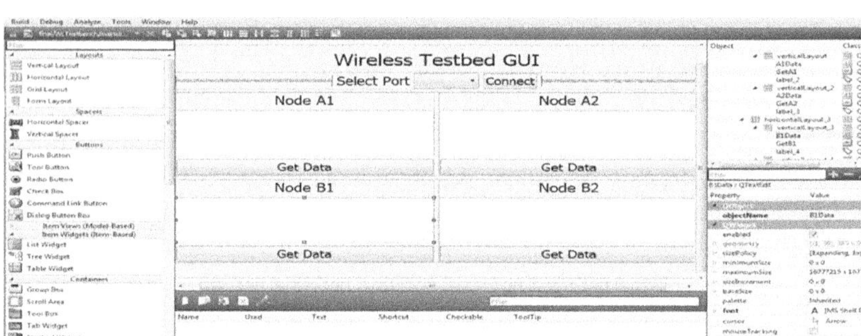

Figure 7.5 GUI in the QT creator.

7.7.1 Wireless network implementation

The network's wireless communication and routing protocols were tested with the QualNet Simulator used for designing and testing, as demonstrated in Figure 7.6. Parameters, such as humidity, rain sensors, light intensity, temperature, and wind direction, were monitored to support crop growth.

The hardware testing of the three PANs with mobile sensor nodes is demonstrated in Figures 7.7 and 7.8 to assess the efficiency of the routing protocol.

The performance of the WSN was evaluated in a heavy-rain zone, demonstrating its performance.

7.7.2 Design and development of the WSN protocol

A new routing protocol is introduced that combines the aspects of Bellman–Ford and Leech routing protocols for efficient data transmission and communication within a WSN.

7.7.3 Optimizing network performance using Bellman–Ford and Leach protocols

The integration of the Bellman–Ford algorithm and Low-Energy Adaptive Clustering Hierarchy (Leach) protocol improves the performance of WSNs in smart agriculture. Each protocol has a different role in optimizing the network performance:

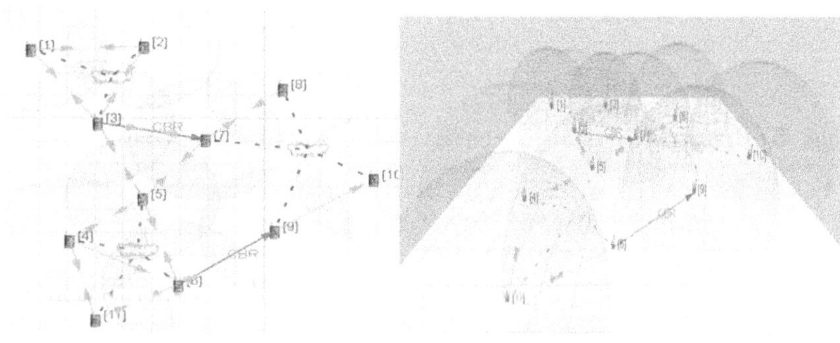

Figure 7.6 Three PAN testing on emulation (hardware in loop).

Figure 7.7 PAN testing on the heavy rain zone in emulation and simulation.

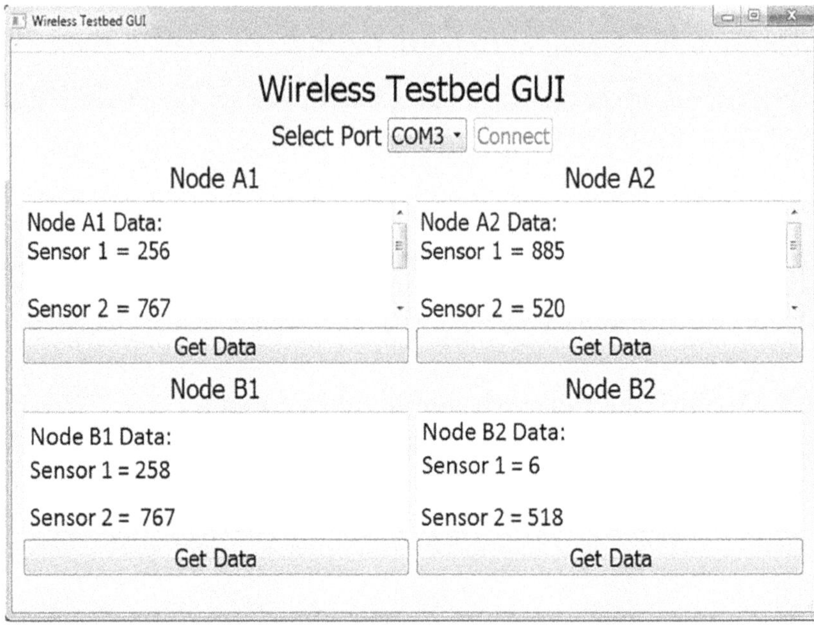

Figure 7.8 Starting window for three PAN.

7.7.4 Bellman–Ford algorithm—efficient data routing

- The Bellman–Ford algorithm determines the shortest path between the sensor nodes and the base station by considering the energy consumption and network distance.
- The purpose is to reduce delays and enhance the efficiency of data transmission so that the transmission of sensor information reaches the processor within a short period.

7.7.5 Leach protocol—energy optimization

- Leach is a cluster-based protocol in which sensor nodes are organized into groups with one designated Cluster Head (CH).
- The CH acquires information from adjacent nodes, processes this information, and transmits the information to the base station to reduce the direct communication load and conserve energy.
- The CHs switch every once in a while to balance the power consumption and prevent any single node from discharging its battery too fast.

When combined with the Leach protocol, the Bellman–Ford algorithm can increase the network efficacy in WSNs, such as precision agriculture applications. With Leach, a few nodes that can directly send data to base stations are used, whereas Bellman–Ford provides optimal paths for routing data transfer. By efficiently distributing energy consumption, this combination prolongs the lifespan of the network, prevents packet loss, leads to better data accuracy, and minimizes latency for faster decision-making. Hence, this hybrid solution enhances the reliability, scalability, and efficiency of WSNs for real-time monitoring and automation along with optimum energy conservation in smart farming.

The mathematical equations implemented in this routing protocol are as follows:

The initial value of the initial sensor node state is defined as $V(x_0)$, the current state $a_t|\epsilon|\mathbb{H}\ x_t$, and the new sensor node state is F (x, a).

$$V(x_0) = \max_{\{a^t\}_{t=0}^{\infty}} \sum_{t=0}^{\infty} \beta^t F\left(x^t, a^t\right) \text{ and } T\ (n)\ =\ \{\frac{\rho}{1} - \rho * \left(mod \frac{1}{\rho}\right) if\ n \in G \quad (7.1)$$

$$\text{where } a_t \in \Gamma\ x_{t+1} = T\left(x^t, a^t\right), \forall t = 0,1,2,3,.. \quad (7.2)$$

$$\max_{a_0} \left\{ F\left(x_{0,}\ a_0\right) + \beta[\ \max_{\{a^t\}_{t=1}^{\infty}} \sum_{t=0}^{\infty} \beta^{t-1} F\left(x_{t,}\ a_t\right) :\in_{x_{t,}}\ \in [\mathrm{T}(\mathrm{n}) = \left\{\frac{\rho}{1} - \rho * \left(mod \frac{1}{\rho}\right)\right] if\ n \in G\ \forall t \geq 1] \right\} \quad (7.3)$$

Recursive functions are written as follows:

$$V(x_0) = \max_{a_0}\{F(x_{0,}a_0) + \beta V\left(x_1\right)\}, where\ a_0 \in \Gamma\left(x_0\right), x_1 = T\left(x_0, a_0\right) \quad (7.4)$$

$$V\left(z\right) = \max_{a \in \Gamma(z)} \{F(x_, a) + \in [\mathrm{T}(\mathrm{n}) = \{\frac{\rho}{1} - \rho * \left(mod \frac{1}{\rho}\right)] + \beta V\left(T\left(x, a\right)\right)\} \quad (7.5)$$

For further calculations, the cost or value of the interval or distance from the sensor node was determined using Equation (7.4). In the case of communication failure by one sensor node, the next value or cost is calculated with reference to Equation (7.5).

7.7.6 Design and development of the geometric algorithm of WSN

Geometric algorithms were implemented and tested in agricultural areas to optimize sensor node communication and network parameters. The AgriWSN algorithm was introduced, which offers improved performance in monitoring agricultural parameters.

The WSN testbed depicted is the agricultural area of the Sangli district, showcasing sensor node placement and communication. A satellite view of the research area in Karanwadi Village, Maharashtra, India, is displayed in Figure 7.9.

7.7 FUZZY LOGIC CONTROLLER

FLC results are showcased in Figures 7.10–7.12, offering optimized control compared to conventional methods, aiding in better decision-making for farming. Fuzzy logic control is discussed as a decision-making tool for agricultural applications that provides approximate solutions to complex problems based on uncertainty. The Mamdani implication method was used to facilitate decision-making in agricultural scenarios.

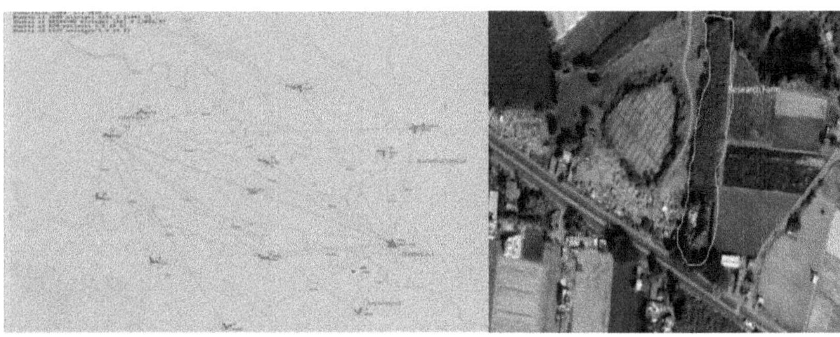

Figure 7.9 Wireless sensor network testbed in the Sangli district and satellite view of research area.

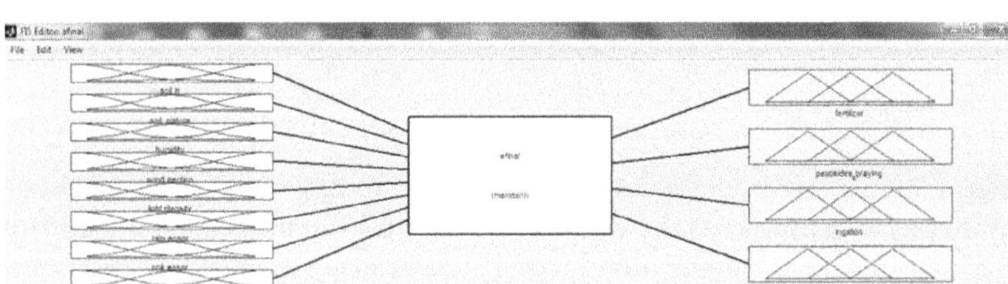

Figure 7.10 Fuzzy logic controller.

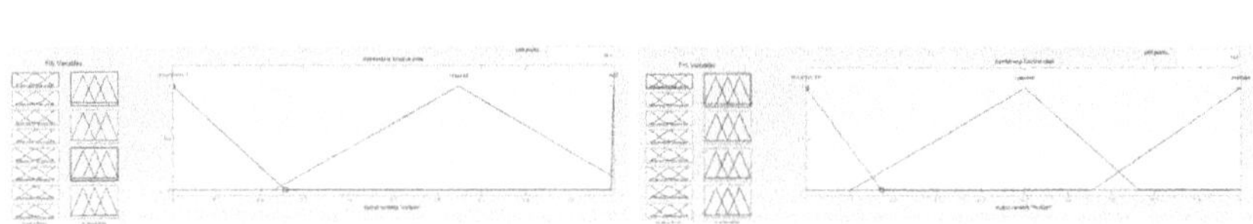

Figure 7.11 Membership value of Agri variables.

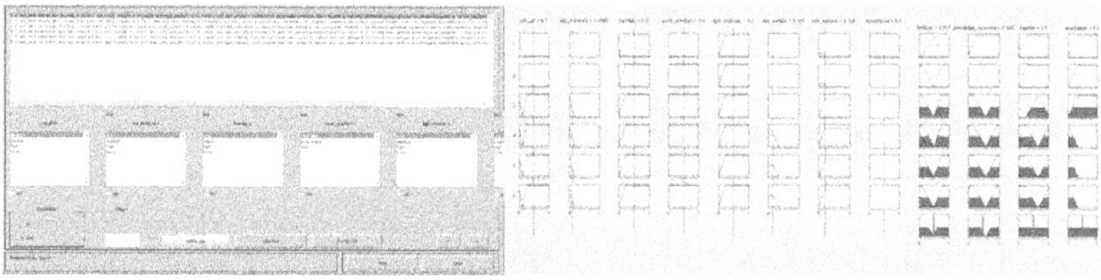

Figure 7.12 Rule setting and viewer in fuzzy logic.

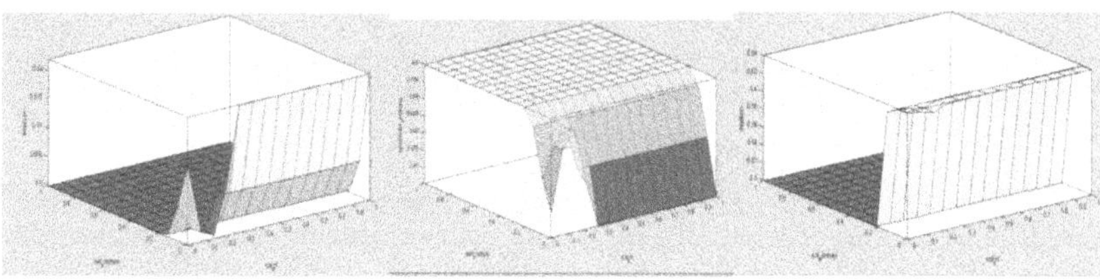

Figure 7.13 Surface viewer of the Agri output.

The design of the FLC is presented, demonstrating its inputs, outputs, and rule-based decision making (Figure 7.13).

7.8 AUTOMATION IN AGRICULTURE

Resource scarcity and pesticide misuse are significant challenges for agriculture, leading to reduced productivity and environmental damage. For instance, studies have shown that over 70% of freshwater is used for irrigation, yet inefficient watering methods result in nearly 50% waste due to evaporation, runoff, or overuse. Similarly, excessive pesticide application leads to soil degradation and harmful residues in crops, with reports indicating that approximately 30% of the global agricultural produce contains unsafe pesticide levels. In India, pesticide overuse has caused severe health issues, and studies have linked it to increased cases of chronic illnesses among farmers. These statistics emphasize the need for intelligent automation, such as WSNs and Fuzzy Logic, to optimize resource usage, minimize waste, and ensure sustainable farming practices.

7.8.1 Impact on the environment and invertebrate non-target organisms of insect-shooting automation in smart agriculture

Automation of pest control as part of precision agriculture through WSN, fuzzy logic, and AI-based systems has greatly increased efficiency. However, environmental impacts must be evaluated, especially in non-target species that are part of the balance for the overall ecology.

Unregulated automated spray pesticides and electrostatic insect traps can injure beneficial insects such as pollinators (pollinating bees and butterflies) and natural predators (ladybugs, spiders, and predacious beetles). With losses in these pest species, pollination efficiency diminishes, which consequently leads to lower crop yields and disrupted ecological balance. Moreover, ongoing spraying with pesticides can bring about chemical deposit build-up in land and water resources, which can spoil soil health and build resistant pest populations in the long term. To alleviate such challenges, the proposed fuzzy logic-driven decision-support system guarantees intelligent pest identification and precise action, thereby avoiding the wasteful use of pesticides. By integrating WSNs for the real-time monitoring of pests, the system can identify harmful pests from beneficial insects to selectively control pests. In addition, implementing environmentally friendly automation techniques, such as biological pest control, pheromone-based insect traps, and AI-driven precision spraying, can assist in reducing the environmental impact while maintaining sustainable agricultural practices.

Automation focuses on managing harmful insects that affect crop health. Research has indicated that certain insects are most active during the evening, specifically from 6:00 p.m. to 9:00 p.m., whereas others are beneficial to crops. To address this, a low-cost insect killer consisting of microfans, plastic or iron tins, blue light LED strips, a controller, and ZigBee modules for communication was engineered. This module is powered by connected solar panels and batteries, automatically activated at 6:00 p.m. and deactivated at 9:00 p.m. Additionally, it offers flexibility in setting activation times based on the farmers' requirements. Harmful insects, beneficial insects, and neutral insects were classified, with insect killers targeting harmful insects and improving crop health.

The insect trapper was maintained on a solar-powered battery. Eighteen insect trappers were maintained on a vegetable farm. On a vegetable farm, the 18 insect traps labored.

7.8.2 Comparing traditional methods of decision-making with fuzzy logic

Fuzzy logic addresses these limitations by accommodating uncertainty and imprecision in the decision-making process. This work presents the principles of fuzzy logic, emphasizing benefits that include translating imprecise inputs for decisions into crisp outputs, determining moderation among variables, and the ability to incorporate expert knowledge from diverse sources. As such, it enables a model that can account for fuzziness and uncertainty in decision-making.

Fuzzy logic, on the other hand, offers a more fluid and intelligent form of decision-making. Rather than restricting itself to mere numerical thresholds, it uses linguistic variables like "low," "moderate," and "high" to facilitate simultaneous analysis of several environmental parameters. For instance, a fuzzy logic-based irrigation system would make a decision on irrigation based on the amalgamation of soil moisture, ambient temperature, humidity, and previous irrigation cycles. This can help the system dynamically adjust water distribution to avoid water wastage while maintaining good soil conditions.

By incorporating sensor information on pest activities, crop health, and weather conditions, fuzzy logic improves the level of precision in pest control. Unlike traditional pesticide spraying methods that run on fixed schedules, fuzzy logic enables real-time decision-making for pest control only when activated, doing away with unnecessary pesticide spraying, and therefore, reduces the environmental impact. A number of studies on applications of precision agriculture and WSN have shown that fuzzy logic improves overall energy efficiency, data accuracy, and automated responses, as compared with conventional approaches. By allowing smooth transitions between different operational states, fuzzy logic enhances system reliability, scalability, and sustainability, making it a powerful tool for modern smart farming.

7.8.3 Wider implications for Indian agriculture: policy and scalability

The study of the integration of WSNs, fuzzy logic, and automation in agriculture is significant for Indian agricultural policies and large-scale applications. Given India's varied agro-climatic regions and limited resources, embracing intelligent automation would enable policymakers to formulate water-saving agricultural policies, precision pest management, and crop monitoring in real time. From a policy standpoint, the government can encourage farmers to implement smart farming technologies through subsidies, training, and infrastructure. Such solutions may be integrated with extant digital projects, such as the National e-Governance Plan for Agriculture (NeGPA), in order to facilitate data-driven decision-making at the national level.

In terms of scalability, this study highlights the viability of rolling out low-cost, energy-efficient sensor networks over small and large farms. With the help of IoT-based farm management systems, farmers will better utilize resources, save operational costs, and maximize productivity. Furthermore, scalable AI-based pest control and irrigation systems can be tailored to various crop types and climatic conditions, thus co-existing with traditional farming and commercial farming practices. Thus, in the long run, the application of these technologies would render Indian agriculture a data-driven, resource-preserving, and eco-friendly enterprise, guaranteeing food security and economic sustainability for millions of farmers.

7.9 CONCLUSION

The research discussed in this paper represents a promising step toward confronting some of the pressing challenges faced by Indian farmers. The intelligent automation system developed, which combines fuzzy logic with WSNs, has provided a crop protection tool to farmers against pests that may cause damage. This, in turn, is expected to help reduce chemical pesticide use. This concept is highly economical, eco-friendly, and farmer-friendly. The adoption of this technology will significantly modernize Indian agriculture and increase the livelihood of farmers nationwide. Future work may involve other areas of application and extension of this resource on a more extended-scale propagation.

REFERENCES

1. A. Bechar and C. Vigneault, "Agriculture robots for field operations: Concepts and components," *Biosystems Engineering*, vol. 149, pp. 94–111, 2016.
2. A. Botta, W. Donato, V. Persico, and A. Pescape, "Integration of cloud computing and internet of things: A survey," *Future Generation Computer Systems*, vol. 56, pp. 684–700, 2016.
3. A. G. Sanchez, F. G. Sanchez, and J. G. Haro, "Towards a new generation of agricultural system data, models and knowledge products: Design and improvement," *Computers and Electronics in Agriculture*, vol. 155, pp. 288–303, 2011.
4. A. A. Kharazian, K. Jamshidi, and M. R. Khayyambashi, "Adaptive clustering in wireless sensor network: Considering nodes with lowest energy," *International Journal of Ad hoc, Sensor & Ubiquitous Computing*, vol. 3, no. 2, pp. 1–12, 2012.
5. A. G. Delavar, S. Shamsi, N. Mirkazemi, and J. Artin, "SLGC: A new cluster routing algorithm in wireless sensor network for decrease energy consumption," *International Journal of Computer Science, Engineering and Applications*, vol. 2, no. 3, pp. 39–51, 2012.
6. B. Sivakumar, P. G. Sekaran, T. S. Prabhu, P. Kumaran, and D. Anandan, "The application of wireless sensor network in the irrigation area automatic system," *International Journal of Computer Technology and Applications*, vol. 3, no. 1, pp. 67–70, 2012.
7. C.-R. Rad, O. Hancu, I. A. Takacs, and G. Olteanu, "Smart monitoring of potato crop: A cyber-physical system architecture model in the field of precision agriculture," *Agriculture and Agricultural Science Procedia*, vol. 6, pp. 73–79, 2015.
8. G. Deepika and P. Rajapirian, "Wireless sensor network in precision agriculture: A survey," in *Proc. IEEE Int. Conf. Emerging Trends in Engineering, Technology, and Science (ICETETS)*, 2016, pp. 1–4.
9. D. Ye, D. Gong, and W. Wang, "Application of wireless sensor networks in environment monitoring," in *Proc. IEEE Power Electronics and Intelligent Transportation System Conf.*, 2009, pp. 205–208.
10. V. Romanov, I., and Galelyuka, Y. Sarakhan, "Wireless sensor networks in agriculture," in *Proc. IEEE 7th Int. Conf. Intelligent Computing and Information Systems (ICICIS'15)*, 2015, pp. 77–80.
11. Y. Dasgupta and G. Darshan, "Application of wireless sensor network in remote monitoring," in *Proc. 1st Int. Conf. Automation, Control, Energy and Systems*, 2014, pp. 1–3.

Chapter 8

Smart farming development integrating IoT-based solutions

Indu

8.1 INTRODUCTION

In the future, smart computing will heavily rely on the Internet of Things (IoT), which is currently driving the transformation of traditional technology into the concept of next-generation everywhere computing. The agricultural sector [1–3] is experiencing notable advancements and product introductions aimed at enhancing productivity and efficiency through IoT technology implementation. Recognized as a crucial sector for global food security, agriculture faces significant challenges in India, including limitations in farm size, technology, trade, government policies, and climate conditions. Various tasks in agriculture, [4–6] such as soil and plant monitoring, environmental factors tracking (moisture and temperature), transportation, supply chain and infrastructure management, control systems, animal monitoring, and pest control, among others, are imperative for productivity enhancement. However, merely monitoring environmental elements does not offer a comprehensive strategy for enhancing agricultural productivity. Additional factors significantly impact productivity, necessitating automation in agriculture to address these challenges comprehensively. Establishing an integrated system that effectively manages all productivity-influencing elements at every level is crucial. Yet, perfect automation in agriculture remains challenging due to various obstacles. While this approach primarily remains in the research domain and is not directly accessible to farmers as a tangible product, this study focuses on developing smart agriculture through IoT devices specifically designed for farmers. Considering India's projected population of approximately 1.5 billion by 2050, with agriculture remaining the primary means of sustenance in rural areas, enhancing agricultural output is paramount. Despite advancements in science and technology, unpredictable power supply or prolonged electrical outages are prevalent. The global use of solar power is increasing as a clean and renewable energy source, with India having significant untapped potential for off-grid solar energy. Agricultural drones are employed to enhance and optimize various farming activities, such as monitoring crops, applying pesticides,

DOI: 10.1201/9781003494782-8

analyzing soil, and creating maps. Drones have become essential tools in the agricultural sector. Farms use unmanned aerial vehicles (UAVs) equipped with sensors and cameras to capture images, create maps, and conduct surveys. There are both ground and aerial drones. Ground drones are autonomous robots with wheels used for surveys and data collection in agricultural areas. Aerial drones, commonly referred to as UAVs or unmanned aircraft systems, are autonomous flying devices.

Drones can be controlled remotely or operate autonomously by following pre-set flight plans stored in their internal systems, which work in conjunction with sensors and GPS technology. These drones provide valuable data on various aspects of agriculture, such as crop health, irrigation, spraying, planting, soil and field conditions, plant counts, and yield prediction. Agricultural surveys can be conducted using drones, either through scheduled services or by keeping them near fields for recharging and maintenance. After the surveys, the drones are transported to nearby labs for data analysis. This approach helps optimize the use of IoT technologies in agriculture. Smart farming leverages advanced technologies like IoT, machine learning, and artificial intelligence to enhance agricultural productivity. These technologies reduce the impact of weather changes on crops through accurate weather predictions, which are essential for crop growth. Crops also face threats from pests and insects, prompting farmers to use chemicals like pesticides and insecticides. However, late detection of pest infestations often leads to significant crop damage. To prevent this, IoT and sensors are implemented in fields to detect pests early and alert farmers, enabling timely intervention and protection.

Water management is another crucial aspect of agriculture. Both overwatering and underwatering can harm crops. Water sensors assist farmers in determining the exact water requirements of their crops, ensuring optimal hydration. Additionally, water prediction technologies help conserve water by preventing wastage. In summary, smart farming integrates advanced technologies to enhance crop production, pest control, and water management. The IoT optimizes farm monitoring by utilizing sensors that detect various activities and send signals to a central device for further processing and decision-making. IoT sensors can monitor leaf moisture, water requirements, and soil moisture. Additionally, IoT sensors are used in livestock management to measure the temperature of animals and their environments.

This project's primary benefit lies in optimizing power usage through efficient water resource management, thereby contributing to the conservation of government-provided free subsidy electricity. This initiative presents a highly effective and cost-efficient irrigation method, ultimately contributing to the automation of the agricultural sector [10,11].

The chapter outlines the structure by discussing IoT, sensors, various types of sensors, the benefits of implementing smart farming, the challenges associated with its development, future prospects, and the conclusion.

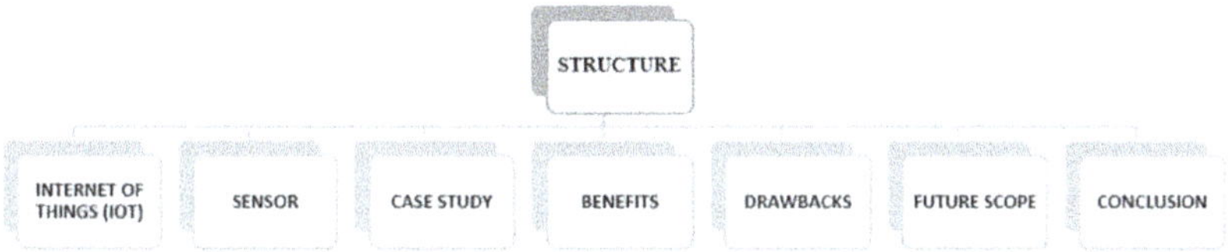

Figure 8.1 The chapter's structure is organized into sections and subsections, providing a detailed explanation of Smart Farming Development through the integration of IoT-based solutions.

Figure 8.1 represents the structure of the chapter, which is divided into subsections such as IoT, sensors, case study, the benefits of implementing smart farming, the challenges associated with its development, future prospects, and the conclusion.

8.2 INTERNET OF THINGS

The IoT refers to a network of interconnected devices that communicate and transfer data to one another using the internet. These devices can transmit data in real time, depending on the internet connection and speed. IoT devices utilize sensors to collect data, such as temperature, and send it to other devices. The primary objective of IoT is to create self-reporting devices capable of real-time communication and data exchange. These devices rely on the internet for data transfer, making the internet essential for IoT functionality.

IoT devices are not limited by area and can easily expand as needed. They enable a wide range of objects to connect to the internet, allowing autonomous operation and real time data transmission. IoT aims to leverage constant connectivity among devices to maximize benefits, offering efficient and reliable solutions for modernizing various sectors, including smart homes, smart cities, industries, and agriculture [7–9].

In agriculture, IoT has demonstrated impressive results in areas like water monitoring and soil pH level measurement. IoT-based solutions in agriculture, known as smart farming, enable automated monitoring and maintenance with minimal human intervention. Smart farming relies on technology instead of human labor, utilizing network technologies such as network architecture, layers, topologies, and protocols.

IoT transforms agriculture by enabling data acquisition from various nodes through remote sensing. IoT devices have unique identities and perform tasks like remote sensing, actuating, and live monitoring of specific data types. These devices exchange data in real time with other connected devices and applications, either directly or indirectly, and can collect, process, and send data to various servers.

The term "Internet" refers to the global communication network that connects billions of computers, facilitating the exchange of information. The IoT is a global network infrastructure capable of self-configuration and operates on standardized and interoperable communication protocols. Within this network, both physical and virtual entities possess unique identities, physical characteristics, and digital personas [10,11]. They use advanced interfaces to seamlessly integrate into the information network, frequently transmitting data about individuals and their environments. IoT technology relies on sensors strategically placed on farms, such as weather station sensors, which collect data and transmit it to analytical tools for detailed analysis.

- **Enhanced Crop Management:** Crop management involves overseeing and optimizing crop production by focusing on factors such as disease detection, water levels, and nutrient content. The crop management team works diligently to maintain the health of the crops using various tools and the latest technologies. IoT devices and sensors play a crucial role in this process by collecting real-time data and transmitting it to other devices. This real-time data collection helps in promptly identifying diseases and addressing them. IoT technology enhances crop management by providing valuable insights into growth patterns, disease detection, and nutrient levels, enabling proactive measures to ensure crop health and increase productivity.
- **Livestock Management:** IoT-enabled devices monitor the health and location of livestock, track feeding patterns, and detect illnesses early. This technology enhances animal welfare, boosts productivity, and minimizes losses. Livestock management focuses on optimizing operations to produce milk, meat, eggs, fish, and other products, while also considering aspects such as chemistry, nutrition, health, and safety.
- **Supply Chain Optimization:** IoT can streamline the agricultural supply chain by providing end-to-end visibility from farm to table, enhancing traceability, reducing food waste, and ensuring fresher produce reaches consumers. An efficient supply chain is crucial for timely transportation, ensuring that production from one city or country is delivered to another on schedule.
- **Predictive Maintenance:** IoT sensors on farming equipment can predict when maintenance is needed, preventing breakdowns and prolonging the life of machinery. This reduces downtime and maintenance costs. Machinery can fail at any time, causing work disruptions. However, IoT sensors continuously monitor and update the machine's condition, ensuring timely maintenance and reducing the likelihood of unexpected failures.
- **Energy Efficiency:** Smart energy management systems monitor and control energy use in farming operations, leading to significant cost savings and reduced environmental impact. Improving energy efficiency can yield benefits for both the environment and the economy.

Lowering costs for the industrial sector enhances competitiveness and job stability. Private households also benefit, particularly those in lower socioeconomic groups, who typically spend a larger portion of their income on energy expenses [12].

- **Sustainability:** By optimizing resource use and reducing waste, IoT contributes to more sustainable agricultural practices. This helps in conserving water, reducing chemical use, and lowering the carbon footprint of farming activities.

8.3 SENSOR

This specific feature sets apart IoT devices from other advanced machines. By incorporating this sensor, the device can recognize the component that has the capability to convert the IoT device from a passive entity into an active and integrated tool. IoT sensors are physical devices that detect changes in an environment and collect data. These components bridge the digital and physical realms within an IoT ecosystem. IoT sensors can measure various parameters such as temperature, pressure, and motion. When connected to a network, they transmit these data for further analysis and action. A sensor is a detection device that can sense environmental changes and convert these sensed data into electrical signals or other necessary output forms according to specific rules. Sensors are responsible for the transmission, processing, storage, display, recording, and control of data. The advancement in IoT has significantly increased the demand for sensors across various sectors, including agriculture and industry. In agriculture, sensors are integral to smart farming, where decisions are increasingly machine-driven rather than human-driven. These sensors measure air temperature, soil pH level, water level, humidity, soil moisture, light intensity, and carbon dioxide levels, collecting data throughout all stages of crop cultivation, from nursery to growth and harvest Conductivity and pH sensors are specifically used to monitor water and fertilizer in agricultural environments. The integrated monitoring system tracks the liquid mixture of fertilizers, ensuring optimal growing conditions.

A single sensor cannot fulfill all the different requirements in agriculture; for example, a sensor measuring soil pH level cannot measure water levels. A crop field is equipped with various types of sensors, which may be connected or operate separately, but they all share data with a central machine (server). This machine processes the data and generates output, guiding further decisions. For instance, if a farmer wants to level a field, they can use a computer-like machine that informs the tractor driver where to operate in real time, showcasing IoT in action.

- **Arduino Mega Microcontroller:** The Arduino microcontroller, based on the Wiring architecture, is an open-source single-board device designed to facilitate the application of electronics across various fields.

It features an Atmel AVR CPU, with software development conducted using a proprietary programming language. Arduino serves as a hardware platform accessible to individuals interested in creating interactive electronic prototypes, employing adaptable and user-friendly hardware and software components. Utilizing the ATMega microcontroller family developed by Atmel as its foundation, there are also individuals and companies producing Arduino replicas using alternative microcontrollers while ensuring hardware compatibility [12–14]. Programs are typically loaded via the bootloader to enhance flexibility; however, bypassing the bootloader and utilizing the downloader allows direct programming of the microcontroller through the ISP port.

- **Soil Moisture:** A soil moisture sensor is a device designed to gauge the moisture content, or water level, present in the soil. Consisting of two probes, this sensor allows electric current to pass through the soil, enabling the measurement of its resistance, which correlates with moisture levels [15,16]. When soil moisture increases, its electrical conductivity rises, leading to lower resistance to electrical current. Conversely, dry soil exhibits high resistance, hindering the flow of electricity. These conditions are relevant for media capable of producing analog voltages with low values. To facilitate microcontroller processing, the voltage is converted into a digital signal.
- **Soil pH Sensor:** A soil pH sensor is a tool employed for gauging the acidity or moisture levels in soil. It covers a pH range from 3.5 to 8. This sensor can directly interface with Arduino or other microcontroller analogue pins, negating the requirement for an intermediary gain module.
- **Anemometer:** The anemometer, commonly known as a wind speed measuring instrument, finds frequent application in meteorology, geophysics, and weather forecasting stations. It serves as a tool for gauging wind speed and other related parameters such as wind pressure, weather conditions, and sea wave height.
- **Real-Time Clock (RTC):** An RTC is a vital component in electronic devices, maintaining accurate time even when the device is powered off or in a low-power state. An RTC actively retains and records date and time information using the I2C data connection protocol, which relies on just two communication lines: SDA and SCL. Embedded systems, which include both hardware and software specifically designed to perform certain tasks, often incorporate RTCs. These systems typically consist of microprocessors/microcontrollers, RAM/ROM, networking elements, I/O units, and storage devices [17,18]. An RTC functions similarly to the brain during sleep—it remains active even when the device is at rest. Just as the brain evaluates the environment and performs essential tasks while we sleep, commanding the body to wake up at the appropriate time, an RTC continues to measure and record time accurately.

An RTC precisely measures and records time within an electronic device, regardless of power status, ensuring continuity even during power loss or energy-conserving “sleep mode.” For instance, in wearable devices like fitness trackers, which eventually enter low-power mode due to finite battery life, an RTC ensures accurate timekeeping. Even with a depleted battery, a fitness tracker can still provide precise time information to the user. An RTC is a low-energy timing solution, ensuring the uninterrupted functionality of key features, even if the device has not been charged for a long time. The IoT represents the convergence of data and internet-connected objects, serving as a fundamental component of the upcoming Internet era. Its primary aim is to automate processes to reduce human intervention. In agriculture, IoT aims to automate many aspects of agricultural operations to improve efficiency and effectiveness [9,19,20].

Traditional farming methods suffer from shortcomings such as inadequate animal management and various inefficiencies, including high labor costs, power consumption, and water usage. These elements are part of traditional farming, as illustrated in Figure 8.2.

This project employs a wireless sensor network to collect data from different sensors and transmit them to a central server using wireless protocols. The gathered data facilitate the acquisition of information on various environmental parameters, enabling comprehensive monitoring of the entire agricultural process [21,22]. However, solely monitoring environmental conditions does not offer an effective strategy for enhancing crop yield, quality, and production. Therefore, the development of a comprehensive and integrated system addressing multiple factors affecting production, including cultivation, harvesting, and post-harvest storage, is imperative.

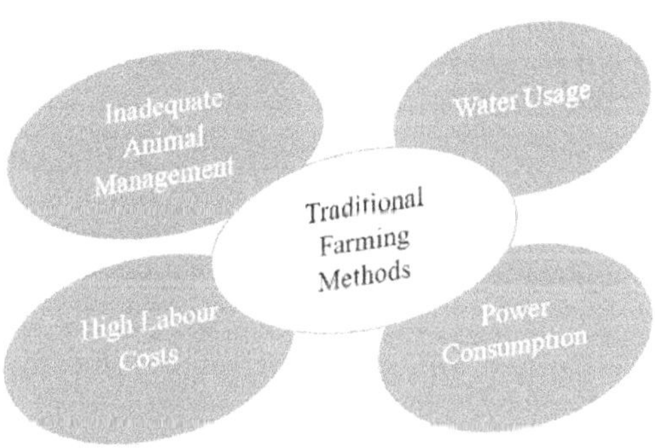

Figure 8.2 Traditional farming encompasses many subsections, such as animal management, and is marked by various inefficiencies, including high labor costs, power consumption, and water usage.

This study introduces a self-autonomous system designed to continuously monitor agricultural fields and execute appropriate controls based on signals received from connected agricultural devices. The system utilizes a GPS-enabled robotic platform capable of performing various tasks such as weeding, sowing, harvesting, and more, both manually and automatically. Additionally, it integrates various types of sensors to measure soil temperature, moisture levels, humidity, growth regulation, and numerous other parameters. Provides a self-sufficient system that can continuously monitor agricultural fields and take necessary actions based on signals from connected agricultural equipment. The system employs a GPS-enabled robotic platform that can be programmed to perform various manual and automated tasks, such as weeding, seeding, harvesting, and more. Additionally, it incorporates multiple sensor types to monitor soil temperature, moisture content, humidity, growth control, and other essential characteristics.

Furthermore, the project includes warehouse management by establishing connections between automated equipment such as balers, tractors, harvesters, and crop planting machines to the internet-connected system [23–25]. The management of these operations will be facilitated by a smart device or computer connected to the internet. The execution of these tasks will be carried out using sensors, Wi-Fi or ZigBee modules, LoRa, cellular connectivity, camera technology, and actuators controlled by a micro-controller and Raspberry Pi.

8.4 CASE STUDY

IoT has emerged as a transformative technology across various fields, including home automation, industry, healthcare, and agriculture. Its impact is particularly significant in agriculture, especially in the production of food and vegetables. However, a major challenge in the widespread adoption of IoT in this sector lies in addressing the immense heterogeneity of the domain. This case study presents an architectural framework designed to model IoT-based systems within the agriculture and food sector. The framework consists of a cohesive set of architectural viewpoints and guidelines for modeling individual IoT-based system architectures. It has been validated through multiple case studies within the European IoF2020 project, encompassing diverse agricultural sub-sectors, conventional and organic farming, early adopters and majority farmers, as well as various roles within the supply chain. The framework proves to be a valuable tool for developing timely, precise, and coherent models of IoT-based system architectures for a wide range of use cases. Additionally, it establishes a common language for aligning system architectures and facilitates the reuse of architectural knowledge across multiple autonomous IoT-based systems in agriculture and food industries. This chapter is centered on two case studies: one focusing on IoT and the other on sensors, highlighting how the integration of IoT and sensors introduces innovative approaches to the agriculture sector. To develop smart farming, IoT sensors play a crucial role by enhancing crop production and minimizing manual labor.

8.5 BENEFITS

Deploying monitoring and data collection systems to oversee soil conditions, air temperature, humidity levels in both air and soil, as well as sunshine intensity across multiple fields, has proven to elevate water utilization efficiency, crop quality, and overall crop yield within extensive agricultural endeavors. Smart farming leverages technologies such as IoT, machine learning, and artificial intelligence to enhance agricultural production. These technologies mitigate the impact of weather changes on crop production by enabling accurate weather predictions, which are crucial for crop growth. Additionally, crops often suffer from pests and insects, leading farmers to use chemicals like pesticides and insecticides. However, farmers sometimes become aware of pest infestations too late, resulting in significant crop damage. To address this issue, IoT and sensors are employed in crop fields to detect pests early and notify farmers, allowing for timely intervention and crop protection.

As the global population continues to grow, the need for increased food production to meet everyone's needs becomes more pressing. Therefore, farming techniques must evolve to effectively meet these demands. As shown in Figure 8.3, incorporating low-cost sensors, data analytics, and IoT platforms can significantly enhance productivity. Other applications of self-autonomous agriculture systems include:

Increase of Production: The adoption of this technology guarantees precision at every stage, including planting, watering, and harvesting, leading to enhanced crop yields.

Conservation of Water: By employing automated and precise crop data, along with moisture and humidity sensors to determine water requirements, water is utilized only when necessary. Consequently, water wastage is mitigated, and conservation efforts are promoted.

Lowered Operation Cost: By employing automated machinery and a comprehensive crop database, the occurrence of human errors is eliminated, thereby avoiding unnecessary expenses associated with inaccuracies. In reality, errors often occur, leading to additional costs for rectification. When humans perform a task, there is a possibility of errors; however, if a machine completes the task, the accuracy is likely to be higher.

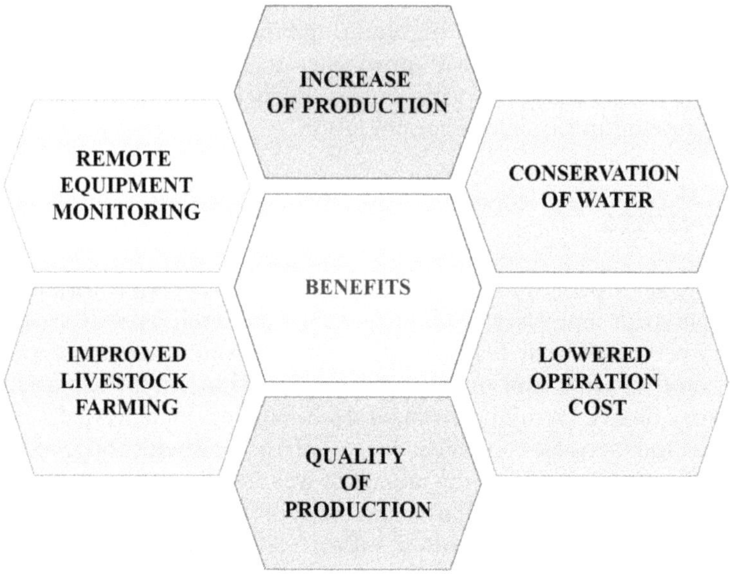

Figure 8.3 Advantages of IoT in agriculture.

Quality of Production: By obtaining precise outcomes, the crop quality is enhanced as pests, soil bacteria, and other detrimental factors are automatically minimized upon detection. Additionally, soil quality is improved, consequently enhancing crop quality [27].

Improved Livestock Farming: Through the employment of sensors and robotics, the health and reproductive patterns of animals are constantly monitored, and automated interventions are initiated as needed, thereby enhancing the management of livestock.

Remote and Equipment Monitoring: Thus, each machine is interconnected, allowing users to operate them manually or utilize pre-programmed algorithms (automation) as desired.

Water management is another critical concern in agriculture. Both excessive and insufficient water supply can harm crops. Water sensors help farmers determine the precise water needs of their crops, ensuring optimal hydration. Moreover, water prediction technologies aid in water conservation by preventing wastage. Overall, smart farming integrates advanced technologies to improve crop production, pest control, and water management. With the global population steadily rising, there's a growing need to boost food production to cater to everyone's needs. Farming practices must evolve to meet these expectations efficiently [25–27]. Implementing surveillance and data-gathering systems to monitor soil properties, air temperature, soil and air humidity, and sunlight intensity across multiple fields has demonstrated improvements in crop quality, yield, and water use efficiency in large-scale farming operations.

8.6 DRAWBACKS

Utilizing IoT for smart agriculture involves leveraging data gathered from various sensors and parameters to make informed decisions and enhance predictions regarding crop productivity and quality. While the intention behind employing IoT in the agricultural sector is commendable, challenges arise in the implementation process. Some drawbacks associated with smart agriculture include:

- A substantial investment is necessary to implement smart agriculture practices in agricultural fields.
- Continuous internet connectivity is essential for utilizing smart agriculture tools; however, internet availability and speed are often limited, especially in rural and remote areas.
- Climate conditions may not always be accurately predicted due to the complexities involved in using smart agriculture technologies.
- The complexity of the technology makes it challenging for uneducated individuals to understand and utilize effectively, limiting access and adoption, particularly among rural populations.
- Collecting vast amounts of heterogeneous data generated by IoT sensors and employing them for numerous agricultural activities, including crop studies, for a large-scale implementation.
- Incorporating a wide range of commercially accessible sensors, weather stations, and cameras, it also supports the integration and utilization of any IoT device, enabling a bring-your-own sensor model. This approach empowers farmers to leverage more advanced IoT sensors according to their individual preferences and financial constraints.
- Incorporating historical crop performance data generated from previous studies alongside diverse data collected from a wide array of IoT devices.

The IoT represents a convergence of data and interconnected devices, poised to become integral to the future Internet. Its core objective is to streamline processes by reducing human intervention. In agriculture, IoT initiatives primarily aim to automate diverse aspects of farming practices for enhanced efficiency and productivity.

8.7 FUTURE SCOPE

The adoption of self-autonomous agriculture systems is strongly recommended and proven to be highly effective, particularly when considering long-term objectives such as enhanced production and automation. By employing self-autonomous systems, the data collected during the process become highly efficient, aiding in field improvement and potentially preventing financial crises among farmers, thereby reducing the incidence of suicides. Moreover, it leads to an increase in both the quantity and quality of agricultural products. Additionally, due to the system's automated nature, labor costs are reduced, and human intervention is minimized. This allows for better measurement of variations within a field and enables farmers to adjust strategies accordingly, resulting in more effective and selective use of pesticides and fertilizers. By incorporating a wide array of commercially available sensors, weather stations, and cameras, the system also supports the integration and use of any IoT device, enabling a bring-your-own-sensor model. This approach allows farmers to utilize more advanced IoT sensors based on their personal preferences and budget limitations. Smart farming represents the future of agriculture, as the global population continues to grow rapidly. Each country must meet its food requirements despite limited land availability, necessitating an increase in food production. Traditional farming methods are no longer sufficient for this purpose, making it essential to transition to smart farming. This advanced, technology-based approach leverages IoT, sensors, and smart machines. These smart machines can detect field needs such as water and pesticide requirements, thereby optimizing the farming process.

8.8 CONCLUSION

IoT devices not only enhance crop yield but also improve cost-efficiency and reduce waste. This chapter aims to introduce a smart IoT-based agriculture stick, allowing farmers to access real-time data such as weather conditions, water levels, soil pH, temperature, soil moisture, and other relevant parameters. IoT is used in smart farming to automate tasks with intelligent devices, minimizing human involvement. This enables efficient environmental monitoring, empowering farmers to adopt smart farming practices and improve overall crop production and quality. The proposed agriculture stick integrates Arduino technology and a breadboard, along with various sensors. The live data feed can be conveniently accessed via mobile phones. The IoT applications deployed in the farming and agriculture sector are proving invaluable in enabling farmers to gather highly valuable data. Given the myriad benefits of integrating IoT into agriculture, it is imperative for farmers to grasp and implement this technology to enhance yield and production. With the global population on the rise, there is a pressing need to produce large quantities of high-quality crops, a goal that can be effectively achieved through the strategic deployment of IoT devices. The primary aim of this paper is to underscore the immense potential of IoT in significantly augmenting information accessibility while concurrently addressing numerous challenges encountered in the agriculture sector.

REFERENCES

1. Baghel, A. S., Bhardwaj, A., & Ibrahim, W. Optimization of pesticides spray on crops in agriculture using machine learning. *Computational Intelligence and Neuroscience*, 2022, 9408535, 2022. Doi: 10.1155/2022/9408535
2. Stočes, M., Vaněk, J., Masner, J., & Pavlík, J. Internet of things (IoT) in agriculture-selected aspects. *Agris On-line Papers in Economics and Informatics*, 8(1), 83–88, 2016.
3. Shenoy, J., & Pingle, Y. IOT in agriculture. In *2016 3rd International Conference on Computing for Sustainable Global Development (INDIACom)* (pp. 1456–1458). IEEE, 2016.
4. Kumar, U., Shukla, S., Malik, I., Singh, S., Bhardwaj, H., Sakalle, A., & Bhardwaj, A. Image encryption using chaotic neural network. In *2021 3rd International Conference on Advances in Computing, Communication Control and Networking (ICAC3N)* (pp. 1424–1427). IEEE, 2021.
5. Sung, T.-W., Tsai, P.-W., Gaber, T., Lee, C.-Y. "Artificial intelligence of things (AIoT) technologies and applications", *Wireless Communications and Mobile Computing*, vol. 2021, Article ID 9781271, 2 pages, 2021. https://doi.org/10.1155/2021/9781271.
6. Muangprathub, J., Boonnam, N., Kajornkasirat, S., Lekbangpong, N., Wanichsombat, A., & Nillaor, P. IoT and agriculture data analysis for smart farm. *Computers and Electronics in Agriculture*, 156, 467–474, 2019.
7. Malik, I., Bhardwaj, A., Bhardwaj, H., & Sakalle, A. IoT-enabled smart homes: Architecture, challenges, and issues. In *Revolutionizing Industrial Automation Through the Convergence of Artificial Intelligence and the Internet of Things* (pp. 160–176). IGI Global, 2023.
8. Mohanraj, I., Ashokumar, K., & Naren, J. Field monitoring and automation using IOT in agriculture domain. *Procedia Computer Science*, 93, 931–939, 2016.
9. Malik, I., & Tarar, S. Cloud-based smart city using internet of things. In *Integration and Implementation of the Internet of Things through Cloud Computing* (pp. 133–154). IGI Global, 2021.
10. Baghel, A. S. Evaluate the growing demand for and adverse effects of pesticides and insecticides on non-target organisms using machine learning. In *2022 6th International Conference on Computing, Communication, Control and Automation (ICCUBEA* (pp. 1–5). IEEE, 2022.
11. Arora, S., Gupta, S., Verma, S., & Malik, I. Prediction of DNA interacting residues. In *2023 International Conference on Computational Intelligence, Communication Technology and Networking (CICTN)* (pp. 54–57). IEEE, 2023.
12. Tzounis, A., Katsoulas, N., Bartzanas, T., & Kittas, C. Internet of Things in agriculture, recent advances and future challenges. *Biosystems Engineering*, 164, 31–48, 2017.
13. Farooq, M. S., & Akram, S. IoT in agriculture:Challenges and opportunities. *Journal of Agricultural Research*, 59(1), 2021, 03681157.
14. Malik, I., Navadia, N. R., Jamshed, A., Verma, L., Singh, T., Bhardwaj, H. Effects of SARS-COV-2 on Blood. In: Tiwari, A., Ahuja, K., Yadav, A., Bansal, J.C., Deep, K., Nagar, A.K. (eds) *Soft Computing for Problem Solving. Advances in Intelligent Systems and Computing*, vol 1393. Springer, Singapore, 2021. https://doi.org/10.1007/978-981-16-2712-5_8

15. Soni, J., Kumar, R. S., Raj, P., Malik, I., Bhardwaj, H. and Sakalle, A. E-BookCart: A candy shop for book enthusiast. In *2021 3rd International Conference on Advances in Computing, Communication Control and Networking (ICAC3N)* (pp. 1655–1659). Institute of Electrical, 2021. https://doi.org/10.1109/ICAC3N53548.2021.9725588
16. Brewster, C., Roussaki, I., Kalatzis, N., Doolin, K., & Ellis, K. IoT in agriculture: Designing a Europe-wide large-scale pilot. *IEEE Communications Magazine*, 55(9), 26–33, 2017.
17. Adesta, E. Y. T., Agusman, D., & Avicenna, A. Internet of things (IoT) in agriculture industries. *Indonesian Journal of Electrical Engineering and Informatics (IJEEI)*, 5(4), 376–382, 2017.
18. Nachankar, P. J., Somani, M. G., Singh, D. M., & Katkar, S. N. IOT in agriculture. *Decision Making*, 1(3), 2018.
19. Dokic, K., Blaskovic, L., & Mandusic, D. From machine learning to deep learning in agriculture – The quantitative review of trends. In *IOP Conference Series: Earth and Environmental Science* (Vol. 614, No. 1, p. 012138). IOP Publishing, 2020.
20. Zhu, N., Liu, X., Liu, Z., Hu, K., Wang, Y., Tan, J., ... & Guo, Y. Deep learning for smart agriculture: Concepts, tools, applications, and opportunities. *International Journal of Agricultural and Biological Engineering*, 11(4), 32–44, 2018.
21. Fiehn, H. B., Schiebel, L., Avila, A. F., Miller, B., & Mickelson, A. Smart agriculture system based on deep learning. In *Proceedings of the 2nd International Conference on Smart Digital Environment* (pp. 158–165). Association for Computing Machinery, 2018.
22. Zhu, N., Liu, X., Liu, Z., Hu, K., Wang, Y., Tan, J., ... & Guo, Y. Deep learning for smart agriculture: Concepts, tools, applications, and opportunities. *International Journal of Agricultural and Biological Engineering*, 11(4), 32–44, 2018.
23. Malik, I., & Singh Baghel, A. Elimination of herbicides after the classification of weeds using deep learning. *International Journal of Sensors Wireless Communications and Control*, 13(4), 254–269, 2023.
24. Altalak, M., Ammad uddin, M., Alajmi, A., & Rizg, A. Smart agriculture applications using deep learning technologies: A survey. *Applied Sciences*, 12(12), 5919, 2022.
25. Malik, I., Baghel, A. S., & Bhardwaj, H. Deep learning for sustainable agriculture: Weed classification model to optimize herbicide application. *Journal of Autonomous Intelligence*, 7(5), 2024.
26. Varman, S. A. M., Baskaran, A. R., Aravindh, S., & Prabhu, E. Deep learning and IoT for smart agriculture using WSN. In *2017 IEEE International Conference on Computational Intelligence and Computing Research (ICCIC)* (pp. 1–6). IEEE, 2017.
27. Boukhris, L., Abderrazak, J. B., & Besbes, H. Tailored deep learning based architecture for smart agriculture. In *2020 International Wireless Communications and Mobile Computing (IWCMC)* (pp. 964–969). IEEE, 2020.

Chapter 9

AI-driven optimization of biosurfactant production from agricultural waste

A sustainable solution to pollution and resource utilization

Vandita Sharma

ABBREVIATIONS

AI	Artificial intelligence
ANN	Artificial neural network
CNN	Connected neural network
DRL	Deep reinforcement learning
FAO	Food and Agriculture Organisation
GA	Genetic algorithm
IoT	Internet of Things
ML	Machine learning
RL	Reinforcement learning
RNN	Recurrent neural network
RSM	Response surface methodology
SMART	Specific, Measurable, Achievable, Relevant and Time-bound

9.1 INTRODUCTION

Across the globe, agricultural waste is an active contributor to greenhouse gas emissions. India alone generates a significant amount of agricultural waste annually. It is estimated that out of 500 million tons of agricultural waste generated annually, 92 million tons are burned each year [1]. This waste, when disposed of or burnt, emits greenhouse gases that are an environmental and health hazard [2]. There is a need for an effective waste management strategy to minimize environmental pollution and promote sustainable agricultural practices hand in hand.

Biosurfactants are surface-active compounds produced by microorganisms. They contribute to the growth of plants by increasing the soil's nutrient availability and promoting the growth of beneficial microflora [3]. Agricultural waste can be utilized as a substrate for the production of biosurfactants via solid-state fermentation [4]. Utilization of agricultural waste is not only a great opportunity to mitigate the harmful effects of global warming but also a sustainable step toward valorizing waste into valuable products.

DOI: 10.1201/9781003494782-9

The problem with obtaining a steady and economical biosurfactant production in a bioprocess is the heterogeneous nature of agricultural waste. This problem can be resolved by the use of Artificial Intelligence (AI). There has been a recent increase in the utilization of AI tools for optimization and enhancement of the productivity of bioprocesses. Predetermined rules and/or identification of data patterns enable AI programs to make independent decisions, thus aiding in optimal production in bioprocesses [5].

The chapter focuses on three aspects: one is the utilization of agricultural waste, the second is the production and application of biosurfactant in agriculture, and the third is the utilization of emerging technology, AI, as a tool in order to optimize and increase the efficiency of production for bioprocesses. The idea is to explore if AI as a technology can be used on agricultural waste, as a substrate, to produce biosurfactants that can be in turn used to enhance soil fertility and provide economic benefits and sustainability to Indian agriculture.

9.2 AGRICULTURAL WASTE

9.2.1 Agricultural waste: an environmental hazard

Agriculture is carried out on approximately 38% of the total surface of the earth. As per data given by the Food and Agriculture Organisation (FAO) of the United Nations, India has the largest cropland area in the world [6]. Of the total land in India, 60.05% is used for farming. Wheat and rice are the main foodgrains sown with rice at 41.17% and wheat at 33.53%, respectively, of the total production [7].

In India, 70% of rural households depend primarily on agriculture for livelihood, and there is a 15% share of agriculture in the national gross domestic product [2]. As per the data given by FAO, 82% of farmers in India are small and marginal farmers. The increased pace of globalization and population has put pressure on farmers and the land, as the crops are sown without giving a break or letting the land rejuvenate naturally. The farmers are in a hurry to dispose of on-field crop residue in order to make the field ready for the next crop. The easiest option they have is the burning of on-field residue. This reckless burning of agricultural waste produces greenhouse gas emissions and causes air and soil pollution. It leads to a dip in the air quality index. It also leads to serious health hazards such as compromise in respiratory function [1].

According to data provided by FAO, India is the highest-incidence country in Southeast Asia in terms of crop residue burning [2]. As per a report of the Indian Agricultural Research Institute, around 63.6% of agricultural waste produced from rice fields is burned annually, resulting in 30 billion dollars' worth of health and economic costs [8]. Figure 9.1 summarizes the effect of burning agricultural waste.

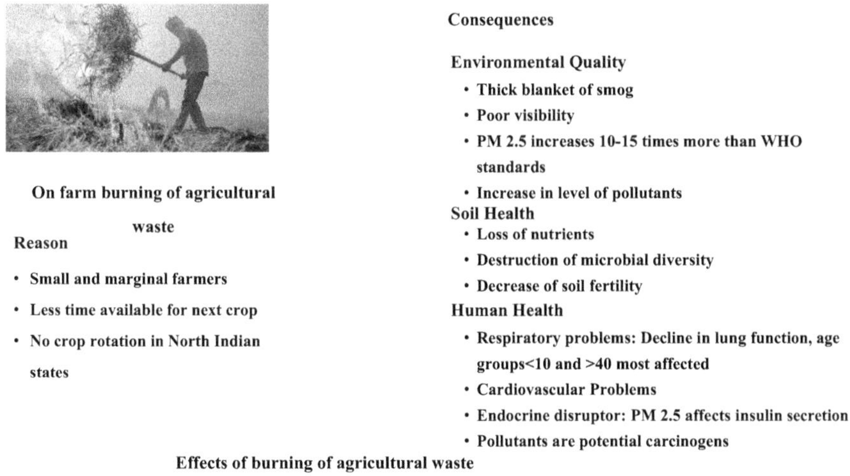

Figure 9.1 A comprehensive understanding of the effect of burning of agricultural waste (Photograph courtesy: PTI).

The factors that lead to the burning of crop residue in India include labor shortage at the time of harvest, preparing the field for sowing the next crop (wheat), use of a combined harvester, and lack of facilities for processing, as most farmers are small and marginal. As a large amount of agricultural waste generated is burnt, it creates an environmental hazard via the emission of greenhouse gases. The burning of agricultural crop residue releases greenhouse gases, air pollutants, and particulate matter, viz, oxides of carbon, oxides of nitrogen, sulfur oxide, methane, and PM 2.5 [2]. A study done by Khan et al. [9] revealed a 50%–75% increment in PM 2.5 and a 40%–45% increase in PM 10 concentration in the months of October and November in the national capital region of Delhi due to the burning of agricultural waste. This not only becomes an environmental hazard but also a health hazard to people in north Indian states. Rice residue, wheat straw, and sugarcane bagasse rate the highest in the emission of pollutants ranging from polyaromatic hydrocarbons to volatile organic compounds [1].

Burning of soil residue negatively impacts soil health as the organic matter of the soil is burnt. According to research done at Punjab Agriculture University every year, 0.824 million tons of nitrogen, phosphorus, and potassium (NPK) are lost from soil due to the burning of agricultural waste. Burning results in an increase in the temperature of the soil, which affects microbial flora in the top 2.5 cm layer of soil. This impacts the natural fertility of soil as beneficial soil microbiota, which contribute to the fixing of important nutrients and promote plant health, are destroyed [1,2].

9.2.2 Type of agricultural waste

Agricultural waste is generated from different sources. Broadly, agricultural waste can be classified as crop residue, agriculture product-based industry, livestock, and aquaculture [10]. Agricultural waste has also been classified as field waste and process waste [11]. Field waste includes stems, stalks, leaves, and seed pods, which are left in the field after the crop has been harvested. The process residues, on the other hand, include husk, bagasse, molasses, and so on, which are generated during the processing step of the crop. Common process residues include wheat straw, rice straw, and bagasse. Amongst the residue generated, rice residue occupies the highest percentage (33%), followed by wheat (21%) and sugarcane residue (17%). Along with on-field agricultural residues, there are plant residues that are obtained after processing in agro-based industries. These include oil cakes like groundnut oil cake, mustard oil cake, sunflower oil cake [12]. Figure 9.2 depicts the percentage of dry biomass contributed by major crops in India [7].

9.2.3 Composition of agricultural waste

Mitchell, Berovic, and Krieger [13] classified the agricultural waste substrate into three groups—starch, lignocellulosic, and substrates containing soluble sugars. These have varying amounts of cellulose, lignocellulose, lignin, and pectin, and consequently different amounts of carbon, nitrogen, micronutrients, and moisture content. As per data given in the review [14], rye straw has the highest lignin content (19.0%–30.8%), and wheat straw has the

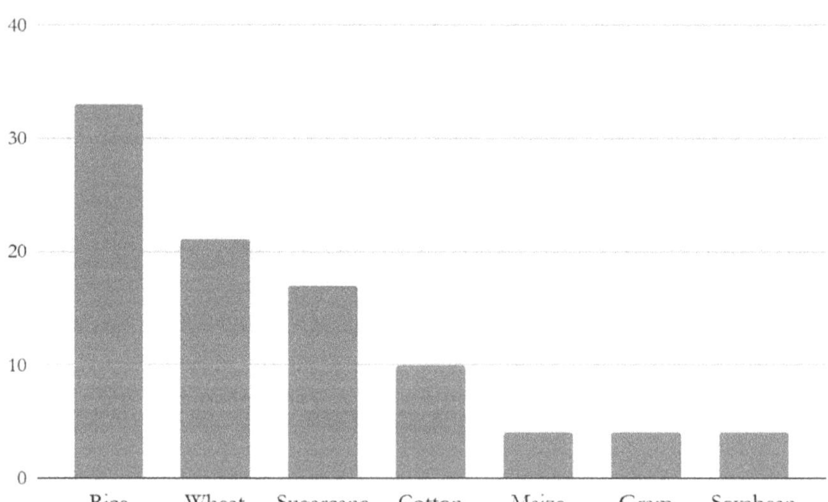

Figure 9.2 Percentage of dry biomass contributed by major crops in India [7].

highest cellulose content (32.9%–49.8%) among the process residues. In the category of food industry processed residues, apple pomace has the highest cellulose content (47.5%). Sugar cane bagasse has a cellulose content of 36.9%–45.7% and lignin content of 18.9%–26.1%, hull-less pumpkin oil cake has the highest protein content (38.3%), followed by Brewer's spent grain (23.4%–27.4%). Oil cakes have a high content of fat [12].

9.2.4 Factors affecting the chemical composition of agricultural waste

As we can understand from the preceding paragraph, the composition of agricultural waste is heterogeneous. The factors affecting the chemical composition of agricultural waste include plant variety, conditions of cultivation, methods of harvesting, conditions of weather, storage conditions, and so forth. Even the analytical methods used for the analysis of individual components affect the understanding of the chemical composition of agricultural waste [14]. With the above variation in composition and constitution, agricultural waste holds the potential to serve as the media constituent for the growth of microorganisms and the production of primary and secondary metabolites [11]. The property of heterogeneity in composition also makes the agricultural waste a suitable candidate for analysis by AI, ensuring an optimum production of the metabolite.

9.2.5 Possibility of using agricultural waste for the production of biosurfactants

Agricultural waste has been utilized as a substrate for the production of bioenergy products such as bioethanol, single-cell protein, food products such as mushrooms, industrial enzymes, and microbial polysaccharides [11,15].

A study by Chinnasamy et al. [16] concludes that agricultural waste, which is an active contributor to greenhouse gas emissions, can also serve as an excellent substrate for the production of biosurfactants. Hence, biosurfactant production using agricultural waste shall not only contribute to carbon neutrality but also generate a product that shall actively contribute to the amelioration of soil fertility issues and enhance crop productivity by encouraging the growth of beneficial microorganisms.

9.3 BIOSURFACTANT

9.3.1 Nature and commercial production

Biosurfactants are low molecular weight surface-active compounds. They are produced by bacteria, yeast, and fungi. Chemically, surfactants are amphiphilic compounds. They have both hydrophilic and hydrophobic moieties that preferentially partition between liquid interfaces with different degrees of polarity and hydrogen bridges, such as oil, water, or air–water interfaces. Surfactants increase the solubility of hydrophobic molecules. This results in a reduction of both surface and interfacial tensions at the oil and water interface [4].

In recent times, microbial surfactants' demand has increased due to the following features: (a) Their diversity and environment-friendly nature, (b) the possibility of their production through fermentation, (c) their greater selectivity and production from renewable sources, (d) low toxicity, (e) their ability to perform function in a wide ranges of pH, temperature, and salinity conditions, (f) their potential application in the environmental protection, (g) crude oil recovery, and (h) their industrial application in health care and food processing. Biosurfactants produced by bacteria are of interest because of their antibacterial, antifungal, and antiviral properties, holding therapeutic and biomedical potential [3]. This may have advantages over their synthetic counterparts as the biosurfactants are environmentally friendly.

According to the work of Merchant and Chakraborty, microorganisms utilize a set of carbon and energy sources for their growth and production of metabolites. The combination of carbon sources with insoluble substrates favors the diffusion of the intracellular medium and the production of different metabolites. Moreover, depending on the nature of the carbon source in the culture medium, the synthesis of biosurfactants can be directed to one of the several metabolic pathways involved in this process. Understanding the above in a simpler way, microorganisms such as bacteria, yeasts, and some filamentous fungi are capable of producing biosurfactants with different molecular structures and surface activities [17].

The commercial production of biosurfactants is a challenge due to the high cost of both raw materials and processing along with low output obtained. Thus, one of the current research challenges includes an increase in the yield along with keeping the cost of raw materials as low as possible. The problem with low-cost raw materials, such as agricultural waste, is the variation in their composition, specifically concerning the carbon and nitrogen ratio in different batches. This variation causes biosurfactant production to become unsteady in industrial scale-ups [18].

9.3.2 Application of biosurfactant in agriculture

Biosurfactants have been studied widely for their application in the field of agriculture. For example, chemical pesticides utilize surfactants as adjuvants. Chemical surfactants are employed for their emulsifying and dispersing properties, which enhance the efficiency of chemical pesticides. Biosurfactants can replace chemical surfactants used in the pesticide industry, thus decreasing soil and groundwater pollution. Research has suggested that biosurfactants can be employed to enhance the quality of agricultural soil. Both organic and inorganic pollutants cause abiotic stress on the crop cultivated in the field.

Biosurfactants accelerate the desorption of hydrophobic moieties, which are bound to particles of soil, aiding in bioremediation [17,19].

Biosurfactants find utilization in the bioremediation of contaminated soil by enhancing the availability and degradation of chemical pollutants and their consequent removal [20]. Biosurfactants also contribute to the protection of plants against pathogens, as they exhibit herbicidal, pesticidal, and antimicrobial activity [21]. Biosurfactants such as sophorolipids and cellobiose lipids have been marginally investigated for their antimicrobial properties against plant pathogens [22,23]. The aforementioned is due to their emulsifying action and their efficiency as dispersal and wetting agents. The antibiofilm activity of a glycolipid biosurfactant has been demonstrated by Patel et al. [24]. Biosurfactant iturin has been shown to activate a plant's defense mechanism by promoting the production of compounds in plants related to defense, thereby enhancing the immunity of plants against external disease-causing agents [25]. Associated with roots and a natural part of the root ecosystem, certain bacterial species such as *Bacillus* have been demonstrated to serve as natural biocontrol agents owing to their production of lipopeptides which act as fungicides [26].

The main highlights of employing biosurfactants as biopesticides in disease management are their environment-friendly nature, high biodegradability, and production from renewable energy resources. Biosurfactants as rhamnolipids and lipopeptides have been extensively studied in the view of crop

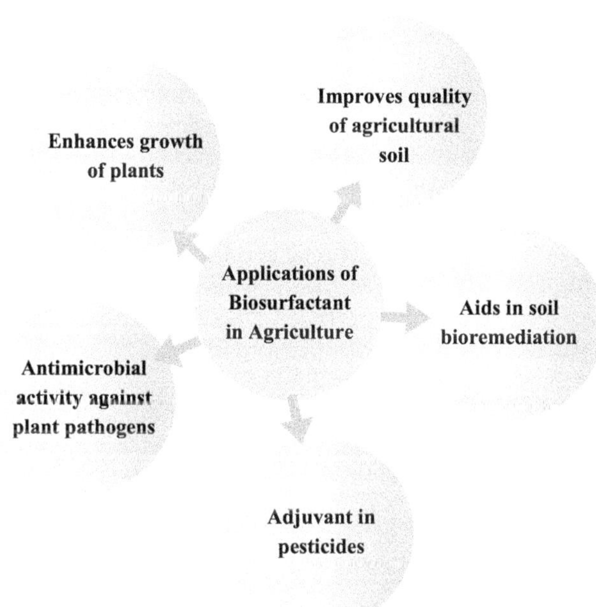

Figure 9.3 Applications of biosurfactants in agriculture.

protection. Chinnasamy et al. [16] submitted in their study that biosurfactant application enhanced the growth of plants. For symbiotic association, biofilm formation is an essential requirement. Biosurfactants have shown to positively influence biofilm formation, thus enhancing the interaction of root nodule-forming bacteria with plants, resulting in plant growth [27].

Biosurfactants help in increasing the rate of composting and enhance the breakdown of organic components, thus making the soil fertile. Biosurfactants have been reported to enhance the interaction of microbes that exhibit beneficial interaction with plants [28]. Thus, biosurfactants have the potential to positively influence the growth of plants overall. Figure 9.3 summarizes the application of biosurfactants in agriculture.

9.4 MICROBIAL PRODUCTION OF BIOSURFACTANTS USING AGRICULTURAL WASTE

9.4.1 Use of complex media for production

Complex media in fermentations are defined as those where the composition of nitrogen and carbon contents, and so forth, varies with each batch. Agricultural waste products are one such kind of complex media. The complex media contain a higher percentage of nitrogen and carbon sources than the minimal media. This leads to higher rates of growth and varying levels of the final product in fermentation [29]. The advantage of using agricultural waste as a raw material for fermentation is that it is a low-cost biomaterial that is being recycled for the generation of useful products. There are a few preliminary steps, viz, pulverization or shredding, which are required. There is minimal requirement and wastage of water in solid-state fermentation in comparison to the requirement in submerged fermentation. The most important point to understand is that, as India is an agricultural country, these raw materials are easily available. Moreover, laboratory-scale research on optimal production parameters enables the grading of the raw materials as those of the highest value [30]. What is meant by the former sentence is that, depending upon various biological parameters (such as the strain of microorganism) and abiotic factors (such as temperature), the researcher can select which agricultural waste gives the highest scale of production. Parameters to be defined for complex media in fermentation include cost efficiency, carbohydrate source addition, and suitability for producing fermentation products under consideration [31].

9.4.2 Solid-state fermentation

Classically, two main types of fermentation processes have been understood based upon the water content required in the process for the metabolic activity of microorganisms—submerged fermentation and solid-state fermentation. Solid-state fermentation is understood as a process where there is an absence or near absence of free water. As the terminology itself states, the raw material or media for fermentation is a non-soluble or solid substrate [18]. Solid-state fermentation has been utilized for a variety of purposes. An important thing to note is that solid-state fermentation is not a new way of production. Classically, vinegar, lactic acid fermentations, mushroom production, and so forth have been carried out using solid-state fermentation processes [11]. Biotechnological fermentation processes utilizing agricultural waste products have found widespread application in the field of bioenergy. The emphasis on bioenergy, viz, bioethanol production using agricultural waste, has been gaining momentum in different countries due to environmental concerns. Overall, the cost of production is much less in the case of solid-state fermentation than in submerged fermentation [11]. There is a lower economic cost as the bioreactor size is smaller, downstream processing cost is reduced, and there is a less chance of contamination as compared to submerged fermentation, hence the sterilization cost is less [32].

Both bacteria, yeasts, and filamentous fungi have been reported to produce biosurfactants on solid substrates. However, yeasts and fungi are more amenable to solid-state fermentation as compared to bacteria. The filamentous fungi with their interwoven mycelial mat are able to procure nutrients by secretion of hydrolytic enzymes from heterogeneously distributed nutrients in solid substrate [33].

Various abiotic factors affecting solid-state fermentation include pH, temperature, moisture, and medium composition. An optimum pH is required for the production and secretion of any secondary metabolite. For yeast and filamentous fungi, the pH of the media is acidic to slightly acidic (ranging from 4.5 to 6), whereas for bacteria, it is slightly alkaline (around 8) [4]. pH also varies according to the genus and/or strain employed or genetically modified for optimal production. Hence, there is no uniform pH at which the production will take place. A specific difference between solid-state and submerged fermentation is the fact that due to a higher quantity of water, the kinetic parameters can be easily monitored and controlled in submerged fermentation [30]. This can also be understood in another way that in solid-state fermentation, there is an issue of thermal conductivity and elevation of temperatures in the inner layers of the solid substrate that can lead to a decrease in productivity as biological processes are temperature-sensitive [13]. As in the case of pH and temperature, filamentous fungi, yeasts, and bacteria all have different optimal temperatures of production, which also vary according to the genus and strain of microorganism utilized for production. Filamentous fungi and yeasts have a requirement of optimal temperature ranging from 25°C to 30°C for optimal biosurfactant production, whereas bacteria require 30°C to 37°C for optimal biosurfactant production. It is also interesting to note that different types of biosurfactants have different optimal temperatures for production, as reported for rhamnolipid and lipopeptide by different research groups [34–37]. Water content and its availability are essential for the survival and growth of the microorganism producing biosurfactants. It also forms a critical factor in enhancing production. Bacteria have a higher water requirement than filamentous fungi or yeast [38]. Biosurfactant-producing microorganisms are isolated naturally from hydrophobic sites like soil or water bodies contaminated with or rich in oil, like oil spill sites. So, it is an obvious understanding that media used for biosurfactant production shall be hydrophobic in nature. However, it has been observed by researchers that the optimal media for biosurfactant production should have both hydrophilic and hydrophobic carbon components [39]. Table 9.1 highlights the different types of agricultural wastes that are employed for the production of biosurfactants [11,14].

Rice straw and sunflower seed meal were employed by Camilios-Neto et al. [40] for the production of 25.5 kg/g dry substrate rhamnolipid. Similarly, El-Housseiny et al. [41] used sugarcane bagasse and sunflower meal to produce 63.3 kg/g dry substrate rhamnolipid.

Usually, agricultural wastes are of different consistency and texture. A good point to be understood here is that as the fermentation progresses, there will be a physical change in the texture and porosity of the solid-state medium. The filamentous fungi, in particular, will make the medium compact. This results in hampering air exchange and temperature distribution, along with changes in mass-energy transfer. Therefore, when solid-state fermentation is taken into production, certain high porosity substrates such as sugarcane bagasse and wheat straw are also employed as supporting agents. The above consideration is of utmost importance as using biosurfactants is an aerobic process [42]. Along with the major requirement of carbon and nitrogen, there is an explicit requirement of micronutrients for the growth of microorganisms and the consequent production of metabolites under consideration [33]. The solid-state

Table 9.1 Different types of agricultural wastes that are employed for the production of biosurfactants

S No	*Example of crop residue*	*Type of crop residue*
1	Oil seed cake	Agro-industry residue
2	Groundnut oil cake	Agro-industry residue
3	Coconut oil cake	Agro-industry residue
4	Gingelly oil cake	Agro-industry residue
5	Palm oil cake	Agro-industry residue
6	Castor oil cake	Agro-industry residue
7	Sunflower oil cake	Agro-industry residue
8	Mahua oil cake	Agro-industry residue
9	Wheat bran	Processed residue
10	Treated molasses and pretreated molasses	Processed residue

substrates provide suboptimal quantities of micronutrients. Therefore, the medium has to be supplied with the required micronutrients as per the composition of the substrate under consideration after extensive research at the laboratory scale. Experiments can be designed to understand the explicit requirement of a particular micronutrient that supports the growth of microorganisms, which is being employed for bioprocesses and to consequently obtain an optimal production [43]. Various reviewers and researchers have elaborated on the advantage of the use of solid-state fermentation in comparison to submerged fermentation. It includes the resistance of microorganisms to catabolite repression in the presence of other carbon sources, economic viability, and reduction in environmental pollution including soil pollution [18].

Different research papers have compared and reported higher production of biosurfactants in solid-state fermentation when compared to submerged fermentation. This result is based on the growth characteristics of the microorganism under study. The exponential phase marks the phase of rapid growth and development. It is in the stationary phase, when the available nutrients become scarce, that the secondary metabolites are produced. Biosurfactants are secondary metabolites. The complexity and heterogeneity of the substrate result in nutrient stress in solid-state fermentation. Research done on iturin A, a biosurfactant that is produced by *Bacillus subtilis*, observed that in solid-state fermentation biofilm is formed, which enhances biosurfactant production in comparison to planktonic growth supported in submerged fermentation [44].

9.4.2.1 Process variables of solid-state fermentation

The success of any bioprocess depends on how well the production parameters affecting the fermentation process can be optimized, making the product commercially feasible. Inoculum size, substrate temperature, and humidity at the start of fermentation are the crucial factors in the production of the biosurfactant [45]. A critical step in solid-state fermentation includes the pretreatment of the substrate. It involves pulverization of the substrate to ensure uniform distribution of matter and energy transfers as much as possible. The substrate is pretreated to hydrolyze the complex components, such as lignin and cellulose, so that simpler carbohydrates are available to biosurfactant producing microorganism. An important point to keep in mind is the diversity evident in different agricultural waste products, due to which an appropriate pretreatment step is to be ascertained after chemical analysis, and the right microbial strain is chosen for the fermentation process. The pretreatment increases the overall efficiency of the process by increasing the production percentage. This is followed by fermentation and finally downstream processing to recover the biosurfactant [11,14].

9.5 ARTIFICIAL INTELLIGENCE

9.5.1 What is AI?

AI is a term that implies the use of algorithms for modeling intelligent behavior with the minimum requirement of human intervention [46]. AI involves the process of perceiving, synthesizing, and inferring information from a given set of data. AI combines machine learning (ML) algorithms with data analytics to enable machines to make decisions (autonomously or semi-autonomously), which would normally require human intervention [47].

AI finds application in the field of biology and its applied fields for drug discovery, understanding host-pathogen interactions, epidemiology research, and so forth. In the field of biotechnology, AI is being utilized for analysis of the structure of biomolecules, understanding metabolic pathways, gene editing, and so forth. In industrial fermentations, AI can help in microbial strain improvement, thus enhancing the quality and quantity of the industrial product. The aforementioned also includes the prediction of optimum conditions for the growth of microbial strains under production and optimization of media used in fermentation [47].

9.5.2 Use of AI in the bioprocess

Increasing the production of any bioprocess for large-scale production involves focusing on increasing production and decreasing production costs. Where agricultural products on the one hand provide a solution to decrease the production cost, the production parameters can be optimized using algorithms and tools that AI offers.

AI programs can make decisions independently through predetermined rules or the identification obtained data patterns [5]. Basic steps in bioprocess include selection of strain, optimization of bioprocess conditions and associated parameters, and scale-up of the process for industrial production and downstream processing. There has been a recent increase in the utilization of AI tools for optimization and enhancement of the productivity of bioprocesses. This encompasses the enhancement of parameters such as biomass, product yield, and the prediction and control of critical process parameters such as aeration (specifically for anaerobic or facultative processes) [48].

The subsequent paragraphs discuss the application of ML in bioprocess development. Figure 9.4 summarizes the SMART solution for bioprocess optimization using agricultural waste.

ML is a subset of AI. Algorithms of ML allow for the application to learn from experience. It is of particular note that a large amount of data are being generated by researchers and scientists in various institutions around the world. These data can be in unstructured form, semi-structured, and or structured [49]. Algorithms of ML are used to train the system for

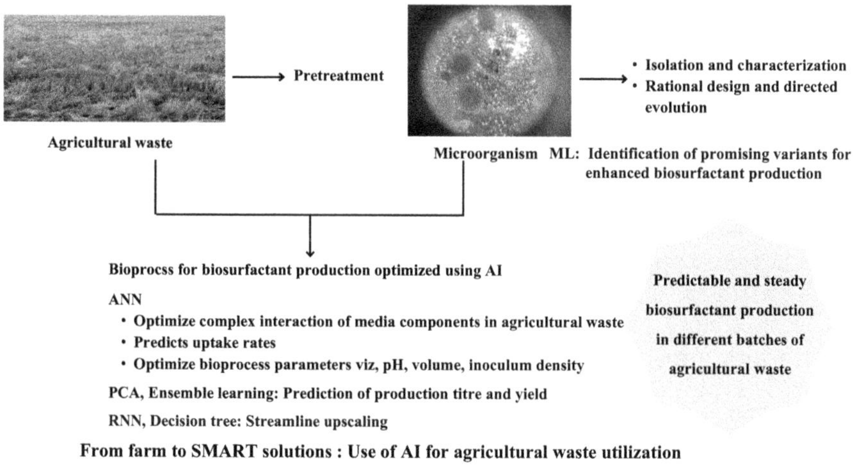

Figure 9.4 An overview of the use of artificial intelligence for biosurfactant production from agricultural waste (Agriculture waste photo courtesy: India Today)

intelligent applications using the considerable insights available from these data. Thus, it allows learning to optimize various stages of bioprocesses from experience.

The very first step in bioprocess engineering for biosurfactant production involves the selection of microbial strain that exhibits the property of quantitative commercial production as well as its qualitative ability to utilize the available agricultural raw material.

The ML approach, on the other hand, is a data-driven approach. It works on identifying patterns in the dataset available and understanding the patterns for similar inputs. It helps in identifying promising variants based on patterns in the collected data. Thus, ML can be used hand in hand with rational design and directed evolution techniques [50]. The development of bioprocesses has traditionally utilized stoichiometric models as well as kinetic genome-scale models [51–53]. Constraint-based modeling has been utilized for making quantitative predictions for bioprocess [54,55]. Data-driven algorithms of ML, on the other hand, provide analysis of complex datasets that can be generated in high throughput output [56]. ML not only complements the traditional modeling pipeline, but also provides hybrid modeling approaches that have the potential to replace mechanistic, genome-scale models completely. Researchers have combined simulated data from genome-scale models with traditionally curated data from different works to predict the production viability for different strains under study. Different ML methods and strategies such as principal component analysis and ensemble learning are applied to get predictions of production titers and yield under different raw material requirements and production conditions. Developed AI models can be used to predict the combination of genes and their respective promoters for the enhancement of production at the genetic level [57]. In addition to this, Artificial Metabolic Network is used to predict uptake rates based on external concentration [58]. The REKINDLE (REconstruction of KINetic models using Deep LEarning) framework has been suggested in a study. In this model, Generative Adversarial Networks are used to obtain mechanistic and kinetic models with biologically feasible dynamics [59]. This model helps in overcoming the challenge of lag in kinetic parameters in experimental data and hence helps in understanding experimental physiology in a better manner.

The artificial neural network (ANN) is an ML-based modeling technique. Outputs for a new data set are predicted by building mathematical models after processing biological information [60]. ANNs have been recognized as tools for modeling and predicting complex non-linear processes. The structure of ANNs is comprised of neurons, which are clusters of processing components closely linked to each other. Input and output layers comprise the dependent and independent variables respectively. Hidden layers in ANN connect the input and output layers, respectively. Functionally, ANNs predict the process more precisely in comparison to the response surface methodology, which is a statistical strategy employed to optimize a bioprocess. ANNs have been utilized for optimizing diverse applications in bioprocess, including incubation temperature, fermentation time, pH, medium volume, inoculum density, and, most importantly, complex interactions of media components. This application becomes relevant when bioprocess deals with undefined media like agricultural waste products whose composition varies from one batch to another [61].

Some examples of the utilization of AI for optimizing bioprocesses are as follows:

- Saini et al. [62] used experimental datasets to train a CNN to optimize regulatory growth factors for increasing biomass production.
- Czinkóczky et al. [63] employed three parameters to predict cell growth and product formation using the built ANN model. They also concluded that harvesting time can be estimated by using an ANN.
- Núñez et al. [64] employed ANN and GA to enhance the modeling accuracy for enzyme production, surpassing the performance of conventional response surface methodology (RSM).
- Ahmad et al. [65] employed ANN for the estimation of the yield of biosurfactant from *Klebsiella sps.* Along with it, the ANN was also used for the estimation of the reduction in surface tension and emulsification index. They used input values as fermentation conditions as temperature, pH, period of incubation, and media components as carbon, nitrogen, and hydrocarbon. Researchers found that the prediction was very similar to the estimation done by the model.
- Jia et al. [66] used online sensor data such as pH and oxidation-reduction potential to develop a deep learning model that can predict

state and performance variables in a dynamic bioprocess. This helps achieve a steady state and optimal production in different batches of bioprocesses.

After having set the parameters for production at laboratory scale, comes the challenge of scaling up the process for industrial fermentations. The production has to deal with the variation in materials and feed streams, complex hydrodynamics, and decreased spatial homogeneity [67]. ML approaches such as Recurrent Neural Networks, decision trees, as well as hybrid models help streamline different contributing parameters for upscaling the industrial production process [68].

Biological processes come with their level of uncertainty; training the system with the available data can help predict the optimal yield and production rate over a long period, thereby enhancing bioproduct formation. Lin et al. [69] have analyzed the integration of transfer learning and deep reinforcement learning (DRL) to enhance bioprocess by improving efficiency and decision-making. The synergy is beneficial as the data obtained from biosurfactant-producing microorganisms can inform the control strategies for new cultures or cultures under investigation. The integration shall also lead to optimizing conditions when co-cultures are employed, as well as predicting metabolic flux.

A step-wise understanding of the process utilizing AI for solid-state fermentation is as follows. First, there shall be initialization of training data, followed by training of the neural network. The fermentation process shall be started thereafter. With the aid of particle swarm optimization, external input parameters shall be solved. Thereafter, the neural network is trained with the help of real-time data. If required, sampling measurement is done spectrophotometrically to analyze the state of growth of microbial culture. The measured data is then used to train the neural network. The analysis shall help in determining if the fermentation process is close to the end. If the culture conditions are predicting the end of fermentation (by spore formation, etc.), the cycle is stopped, or else the external input parameters are again determined. In the above manner, production efficiency shall be improved [70].

Chemically, biosurfactants are of different types, viz, glycolipid and lipopeptide [71]. AI plays a crucial role in enhancing the accuracy of determining the chemical structure of biosurfactants. AI algorithms, such as neural networks and fuzzy inference systems, have been utilized to model enzymatic esterification reactions of fatty acids with sugars, improving predictive capabilities with high accuracy [72].

It can be well understood from the above literature review that AI and ML tools help in the estimation and prediction of biosurfactant production. Using these tools, agricultural waste can be efficiently utilized and developed for biosurfactant production economically. The relationship and variability in different parameters can be streamlined by training the system, in several experimental runs, with variations in agricultural waste. This helps in modeling and analyzing complex relationships between inputs and variables, which shall result in optimizing the biosurfactant quality and shall offer predictions for large-scale agricultural waste utilization.

9.6 CASE STUDY: AI-DRIVEN OPTIMIZATION OF BIOSURFACTANT PRODUCTION FROM AGRICULTURAL WASTE IN INDIA AND THIRD-WORLD COUNTRIES

9.6.1 Background and context

In India and many other third-world countries, agricultural waste is abundant due to the vast agricultural activities undertaken in these regions. However, this waste often remains underutilized, contributing to environmental pollution and lost economic potential. Biosurfactants offer an environmentally friendly alternative to synthetic surfactants and have significant potential for agriculture, industrial, pharmaceutical, and environmental applications.

9.6.2 Objectives

This case study explores the potential of AI-driven optimization for biosurfactant production from agricultural waste in India and other third-world countries. It addresses the following objectives:

1. **Sustainable Resource Utilization:** Transforming agricultural waste into a valuable product—biosurfactant.
2. **Pollution Reduction:** Addressing the issue of agricultural waste management and environmental pollution issues.
3. **Economic Viability:** Reducing costs and increasing scalability for biosurfactant production.

9.6.3 AI-driven solutions

1. **Agricultural Waste Selection and Utilization:** A substantial amount and variety of agricultural waste, including rice husks, sugarcane bagasse, and wheat straw, is generated. ANN and statistical optimization strategies can be employed to predict and optimize the use of agricultural waste as substrates for biosurfactant production, resulting in maximization of yield.

   Example: In a study by Sharon et al. [73], ANNs have been employed to predict biosurfactant production by utilizing dependent variables of the bioprocess. By analyzing different factors, ANN can help identify which agricultural waste or which batch would yield the most biosurfactant, thereby enhancing the efficiency and cost-effectiveness of production. Thus, AI tools can guide local farmers and industries to optimize agricultural waste usage.

2. **Process Optimization Using AI:** The process of biosurfactant production can be optimized using AI tools such as ANN, GA, and RL.

   Example: Training the model with experimental data on fermentation conditions can result in predicting the best conditions for maximum yield. This method can be applied to local production processes, significantly reducing production time and increasing output. The traditional optimization method as RSM often lacks the accuracy and efficiency that AI tools bring, thus bringing forth the transformative potential of AI [74].
3. **IoT and Smart Sensors for Real-Time Monitoring:** Integrating IoT devices with AI systems can revolutionize biosurfactant production in India and the third-world countries by enabling real-time monitoring and adjustments. Key parameters such as, pH, temperature, oxygen levels, and enzyme dosage can be tracked by smart sensors during bioprocesses, while AI algorithms analyze the available data to automatically optimize the process. The development of such a system would reduce resource wastage, energy consumption, and the need for manual interventions.

   Example: In rural areas, access to high-end laboratory equipment is limited. Low-cost IoT sensors integrated with AI can help small and marginal farmers to optimize their fermentation processes without expensive infrastructure [75].
4. **AI for Sustainability and Waste Valorization:** AI-driven technologies can contribute to the circular economy by turning agricultural waste into high-value product biosurfactants.

   Example: In Maharashtra (India), sugarcane waste (bagasse) is abundant, but it is often underutilized. Similar is the case with countries like Sudan [76], where waste from the sugar industry has immense health impacts. AI systems could help determine the most efficient use of this agricultural waste for biosurfactant production, thus reducing pollution and creating new economic opportunities for farmers. AI-driven analytics help in assessing sustainability metrics, thus ensuring alignment of biosurfactant production with environmental goals [77].

9.6.4 Challenges and barriers

1. **Infrastructure Constraints:** Many regions in India and third-world countries face infrastructural challenges, such as limited access to reliable electricity, internet, and high-tech equipment.
2. **Awareness and Education:** There is a need to build awareness and educate local stakeholders about the potential of AI and biosurfactant production technologies.

3. **Data Availability:** The effectiveness of AI models depends on the availability of high-quality, region-specific data. Data collection and sharing frameworks need to be established to support AI development in these areas.

9.7 CONCLUSION

It is a well-established fact now that agricultural waste management is a global issue and needs to be addressed. The advancement in technology has opened the gates to explore and experiment with varied solutions. Some of the key highlights and benefits of the proposed solution are listed below:

- It processes the application of AI technologies for bioprocess control and its optimization.
- Uses AI to help in the optimization of bioprocess to get steady production from a complex and undefined raw material resource—agricultural waste.

The accurate estimation and prediction of biosurfactant production with AI tools will help decrease the cost of production, enhance production efficiency, optimize effective parameters of biosurfactant production, and aid in the development and utilization of agricultural waste. Conversion of agricultural waste into biosurfactant using AI as the technology will not only help in managing agricultural waste optimally and effectively but will also ameliorate the problem of deteriorating soil health, resulting in better harvest per acre. This shall directly aid in elevating the economic condition of small and marginal farmers in two ways: one, as suppliers of raw material to the manufacturer of biosurfactant, and second, their harvest shall be of a better quality and quantity. Hence, it will make farmers a part of the solution, encouraging them not to burn the agricultural waste on the farm.

Sustainable agriculture solutions obtained through SMART use of technology are the need of the hour, and the efforts of all stakeholders, viz, farmers, industrial organizations, researchers, and AI experts, are required to boost the development of novel methods for sustainable agriculture.

9.8 FUTURE SCOPE

The integration of AI-driven optimization for biosurfactant production holds significant promise, not only in transforming agricultural waste management practices but also in contributing to sustainable economic growth. The potential avenues for future research, development, and application are as discussed below.

AI techniques such as DRL and transfer learning can improve predictive capabilities, handle more complex optimization problems, and improve real-time decision-making in fermentation processes by continuously learning from dynamic environmental conditions. This shall help in anticipating future process needs based on data generated in different batches, providing long-term solutions for biosurfactant production scalability, thus generating robust and adaptable process control systems. Integration of multi-omics data with AI can help in mapping metabolic pathways of biosurfactant production via the utilization of different substrates by different organisms. This shall help in the development of engineered strains with enhanced surfactant production. Collaborative AI systems can be developed to facilitate the sharing of results from pretreatment processes, production optimization strategies, and biosurfactant production conditions across various agricultural waste types and regions, thereby accelerating advancements in biosurfactant production. Since most farmers in India are small and marginal, a solution derived in Indian conditions shall be applicable in third-world countries as well. So, future research should focus on the development of low-cost AI systems, which are tailored to rural settings. Thus, AI can play a pivotal role in promoting sustainable, cost-effective, and environmentally friendly biosurfactant production practices that contribute to a circular economy and reduce environmental pollution from agricultural waste.

REFERENCES

1. S. Bhuvaneshwari, H. Hettiarachchi, and J. Meegoda, "Crop residue burning in India: Policy challenges and potential solutions," *International Journal of Environmental Research and Public Health*, vol. 16, no. 5, p. 832, Mar. 2019. https://doi.org/10.3390/ijerph16050832.
2. A. Dutta, A. Patra, K. K. Hazra, C. P. Nath, N. Kumar, and A. Rakshit, "A state of the art review in crop residue burning in India: Previous knowledge, present circumstances and future strategies," *Environmental Challenges*, vol. 8, p. 100581, Aug. 2022, https://doi.org/10.1016/j.envc.2022.100581.
3. A. Kumar, S. K. Singh, C. Kant, H. Verma, D. Kumar, P. P. Singh, et al. "Microbial biosurfactant: A new frontier for sustainable agriculture and pharmaceutical industries," *Antioxidants*, vol. 10, no. 9, p. 1472, 2021. https://doi.org/10.3390/antiox10091472.
4. I. M. Banat, Q. Carboué, G. Saucedo-Castañeda, and J. De Jesús Cázares-Marinero, "Biosurfactants: The green generation of speciality chemicals and potential production using Solid-State fermentation (SSF) technology," *Bioresource Technology*, vol. 320, p. 124222, Jan. 2021, https://doi.org/10.1016/j.biortech.2020.124222.
5. J. K. Khanijou et al., "Metabolomics and modelling approaches for systems metabolic engineering," *Metabolic Engineering Communications*, vol. 15, p. e00209, Dec. 2022, https://doi.org/10.1016/j.mec.2022.e00209.

6. FAOSTAT: Land Use," *Food and Agriculture Organization of the United Nations*, https://www.fao.org/faostat/en/#data/RL [Accessed Nov 2025]
7. "Final estimates of production of major crops released for the year 2022–23," https://pib.gov.in/PressReleaseIframePage.aspx?PRID=1968931#:~:text=As%20per%20Final%20Estimates%2C%20the (accessed May 29, 2024).
8. S. Singh, B. R. Kiran, and S. V. Mohan, "Carbon farming: a circular framework to augment CO2 sinks and to combat climate change," *Environmental Science. Advances*, Jan. 2024, https://doi.org/10.1039/d3va00296a.
9. A. A. Khan, K. Garsa, P. Jindal, and P. C. S. Devara, "Effects of stubble burning and firecrackers on the air quality of Delhi," *Environmental Monitoring and Assessment*, vol. 195, no. 10, Sep. 2023, https://doi.org/10.1007/s10661-023-11635-6.
10. B. Koul, M. Yakoob, and M. P. Shah, "Agricultural waste management strategies for environmental sustainability," *Environmental Research*, vol. 206, p. 112285, Oct. 2021, https://doi.org/10.1016/j.envres.2021.112285.
11. P. K. Sadh, S. Duhan, and J. S. Duhan, "Agro-industrial wastes and their utilization using solid state fermentation: A review," *Bioresources and Bioprocessing*, vol. 5, no. 1, Jan. 2018, https://doi.org/10.1186/s40643-017-0187-z.
12. S. Ramachandran, S. K. Singh, C. Larroche, C. R. Soccol, and A. Pandey, "Oil cakes and their biotechnological applications – A review," *Bioresource Technology*, vol. 98, no. 10, pp. 2000–2009, Jul. 2007, https://doi.org/10.1016/j.biortech.2006.08.002.
13. D. A. Mitchell, M. Berovic, and N. Krieger, "Overview of solid state bioprocessing," *Biotechnology Annual Review*, pp. 183–225, Jan. 2002, https://doi.org/10.1016/s1387-2656(02)08009-2.
14. G. Šelo, M. Planinić, M. Tišma, S. Tomas, D. Koceva Komlenić, and A. Bucić-Kojić, "A comprehensive review on valorization of agro-food industrial residues by solid-state fermentation," *Foods*, vol. 10, no. 5, p. 927, Apr. 2021, https://doi.org/10.3390/foods10050927.
15. V. Saravanan and S. Vijayakumar, "Production of biosurfactant by Pseudomonas aeruginosa PB3A using agro-industrial wastes as a carbon source," *Malaysian Journal of Microbiology*, vol. 10, no. 1, pp. 57–62, 2014, https://api.semanticscholar.org/CorpusID:55842421.
16. M. Chinnasamy, R. Rathanasamy, S. Selvam, H. K. Mohankumar, M. Anandraj, and S. K. Pal, "Biosurfactant as biostimulant: Factors responsible for plant growth promotions," in *Applications of Biosurfactant in Agriculture*, Academic Press, 2022, pp. 45–68. https://doi.org/10.1016/b978-0-12-822921-7.00010-6.
17. A. Roy, "A review on the biosurfactants: properties, types and its applications," *Journal of Fundamentals of Renewable Energy and Applications*, vol. 08, no. 01, Jan. 2018, https://doi.org/10.4172/2090-4541.1000248.
18. M. A. Lizardi-Jiménez and R. Hernández-Martínez, "Solid state fermentation (SSF): Diversity of applications to valorize waste and biomass," 3 Biotech, vol. 7, no. 1, Apr. 2017, https://doi.org/10.1007/s13205-017-0692-y.
19. D. P. Sachdev and S. S. Cameotra, "Biosurfactants in agriculture," *Applied Microbiology and Biotechnology*, vol. 97, no. 3, pp. 1005–1016, Jan. 2013, https://doi.org/10.1007/s00253-012-4641-8.
20. A. A. P. S. Filho, A. Converti, R. De Cássia F Soares Da Silva, and L. A. Sarubbo, "Biosurfactants as multifunctional remediation agents of environmental pollutants generated by the petroleum industry," *Energies*, vol. 16, no. 3, p. 1209, Jan. 2023, https://doi.org/10.3390/en16031209.

21. T. G. Ambaye, M. Vaccari, S. Prasad, and S. Rtimi, "Preparation, characterization and application of biosurfactant in various industries: A critical review on progress, challenges and perspectives," *Environmental Technology & Innovation*, vol. 24, p. 102090, Nov. 2021, https://doi.org/10.1016/j.eti.2021.102090.
22. S. Yoshida et al., "Effects of biosurfactants, mannosylerythritol lipids, on the hydrophobicity of solid surfaces and infection behaviours of plant pathogenic fungi," *Journal of Applied Microbiology*, vol. 119, no. 1, pp. 215–224, May 2015, https://doi.org/10.1111/jam.12832.
23. R. O. Penha, L. P. S. Vandenberghe, C. Faulds, V. T. Soccol, and C. R. Soccol, "Bacillus lipopeptides as powerful pest control agents for a more sustainable and healthy agriculture: Recent studies and innovations," *Planta*, vol. 251, no. 3, Feb. 2020, https://doi.org/10.1007/s00425-020-03357-7.
24. M. Patel et al., "Inhibition of bacterial adhesion and antibiofilm activities of a glycolipid biosurfactant from Lactobacillus rhamnosus with its physicochemical and functional properties," *Antibiotics*, vol. 10, no. 12, p. 1546, Dec. 2021, https://doi.org/10.3390/antibiotics10121546.
25. D. A. Yaraguppi, Z. K. Bagewadi, N. R. Patil, and N. Mantri, "Iturin: A promising cyclic lipopeptide with diverse applications," *Biomolecules*, vol. 13, no. 10, p. 1515, Oct. 2023, https://doi.org/10.3390/biom13101515.
26. W. Jumpathong, B. Intra, J. Euanorasetr, and P. Wanapaisan, "Biosurfactant-producing Bacillus velezensis PW192 as an anti-fungal biocontrol agent against Colletotrichum gloeosporioides and Colletotrichum musae," *Microorganisms*, vol. 10, no. 5, p. 1017, May 2022, https://doi.org/10.3390/microorganisms10051017.
27. E. Z. Ron and E. Rosenberg, "Natural roles of biosurfactants," *Environmental Microbiology*, vol.3, no. 4, pp. 229–236, Apr 2001. https://doi.org/10.1046/j.1462-2920.2001.00190.x.
28. V. Nihorimbere, M. O. Marc, M. Smargiassi, and P. Thonart, "Beneficial effect of the rhizosphere microbial community for plant growth and health," *Biotechnology, Agronomy, Society and Environment*, vol. 15, pp. 327–337, 2011.
29. J. Saxena and R. S. Tanner, "Optimization of a corn steep medium for production of ethanol from synthesis gas fermentation by Clostridium ragsdalei," *World Journal of Microbiology & Biotechnology Incorporating the MIRCEN Journal of Applied Microbiology and Biotechnology/World Journal of Microbiology & Biotechnology*, vol. 28, no. 4, pp. 1553–1561, Nov. 2011, https://doi.org/10.1007/s11274-011-0959-0.
30. V. Kumar, V. Ahluwalia, S. Saran, J. Kumar, A. K. Patel, and R. R. Singhania, "Recent developments on solid-state fermentation for production of microbial secondary metabolites. Challenges and solutions," *Bioresource Technology*, vol. 323, p. 124566, Mar. 2021, https://doi.org/10.1016/j.biortech.2020.124566.
31. V. Singh, S. Haque, R. Niwas, A. Srivastava, M. Pasupuleti, and C. K. M. Tripathi, "Strategies for fermentation medium optimization: An in-depth review," *Frontiers in Microbiology*, vol. 7, Jan. 2017, https://doi.org/10.3389/fmicb.2016.02087.
32. C. K. Venil, N. Z. B. Yusof, and W. A. Ahmad, "Solid state fermentation utilizing Agro-Industrial waste for microbial pigment production," in *Sustainable Agriculture towards Food Security*, Springer, 2017, pp. 375–381. https://doi.org/10.1007/978-981-10-6647-4_20.

33. R. King et al. "A conserved fungal glycosyltransferase facilitates pathogenesis of plants by enabling hyphal growth on solid surfaces," *PLOS Pathogens*, vol. 13, no. 10, p. e1006672, Oct. 2017, https://doi.org/10.1371/journal.ppat.1006672.
34. S. Nalini, R. Parthasarathi, and V. Prabudoss, "Production and characterization of lipopeptide from Bacillus cereus SNAU01 under solid state fermentation and its potential application as anti-biofilm agent," *Biocatalysis and Agricultural Biotechnology*, vol. 5, pp. 123–132, Jan. 2016, https://doi.org/10.1016/j.bcab.2016.01.007.
35. P. Jiménez-Peñalver, T. Gea, A. Sánchez, and X. Font, "Production of sophorolipids from winterization oil cake by solid-state fermentation: Optimization, monitoring and effect of mixing," *Biochemical Engineering Journal*, vol. 115, pp. 93–100, Nov. 2016, https://doi.org/10.1016/j.bej.2016.08.006.
36. J. Wu et al "Production of rhamnolipids by semi-solid-state fermentation with Pseudomonas aeruginosa RG18 for heavy metal desorption," *Bioprocess and Biosystems Engineering*, vol. 40, no. 11, pp. 1611–1619, Aug. 2017, https://doi.org/10.1007/s00449-017-1817-8.
37. L. A. Lourenço, M. D. A. Magina, L. B. B. Tavares, S. M. A. G. U. De Souza, M. G. Román, and D. A. Vaz, "Biosurfactant production by Trametes versicolor grown on two-phase olive mill waste in solid-state fermentation," *Environmental Technology*, vol. 39, no. 23, pp. 3066–3076, Sep. 2017, https://doi.org/10.1080/09593330.2017.1374471.
38. Q. He, H. Peng, M. Sheng, S. Hu, J. Qiu, and J. Gu, "Humidity control strategies for solid-state fermentation: Capillary water supply by water-retention materials and negative-pressure auto-controlled irrigation," *Frontiers in Bioengineering and Biotechnology*, vol. 7, Oct. 2019, https://doi.org/10.3389/fbioe.2019.00263.
39. K. S. T. Souza, E. J. Gudiña, R. F. Schwan, L. R. Rodrigues, D. R. Dias, and J. A. Teixeira, "Improvement of biosurfactant production by Wickerhamomyces anomalus CCMA 0358 and its potential application in bioremediation," *Journal of Hazardous Materials*, vol. 346, pp. 152–158, Mar. 2018, https://doi.org/10.1016/j.jhazmat.2017.12.021.
40. D. Camilios-Neto et al "Production of rhamnolipids in solid-state cultivation using a mixture of sugarcane bagasse and corn bran supplemented with glycerol and soybean oil," *Applied Microbiology and Biotechnology*, vol. 89, no. 5, pp. 1395–1403, Nov. 2010, https://doi.org/10.1007/s00253-010-2987-3.
41. G. S. El-Housseiny, K. M. Aboshanab, M. M. Aboulwafa, and N. A. Hassouna, "Rhamnolipid production by a gamma ray-induced Pseudomonas aeruginosa mutant under solid state fermentation," *AMB Express*, vol. 9, no. 1, Jan. 2019, https://doi.org/10.1186/s13568-018-0732-y.
42. Q. Carboué, M. Claeys-Bruno, I. Bombarda, M. Sergent, J. Jolain, and S. Roussos, "Experimental design and solid state fermentation: A holistic approach to improve cultural medium for the production of fungal secondary metabolites," *Chemometrics and Intelligent Laboratory Systems*, vol. 176, pp. 101–107, May 2018, https://doi.org/10.1016/j.chemolab.2018.03.011.
43. C. R. Soccol, E. S. F. Da Costa, L. A. J. Letti, S. G. Karp, A. L. Woiciechowski, and L. P. De Souza Vandenberghe, "Recent developments and innovations in solid state fermentation," *Biotechnology Research & Innovation*, vol. 1, no. 1, pp. 52–71, Jan. 2017, https://doi.org/10.1016/j.biori.2017.01.002.

44. S. Mizumoto, M. Hirai and M. Shoda, "Production of lipopeptide antibiotic iturin A using soybean curd residue cultivated with Bacillus subtilis in solid-state fermentation," *Applied Microbiology and Biotechnology*, 72, 869–875. 2006. https://doi.org/10.1007/s00253-006-0389-3.
45. S. Dabaghi, S. A. Ataei, and A. Taheri, "Production of rhamnolipid biosurfactants in solid-state fermentation: Process optimization and characterization studies," *BMC Biotechnology*, vol. 23, no. 1, Jan. 2023, https://doi.org/10.1186/s12896-022-00772-4.
46. P. Hamet and J. Tremblay, "Artificial intelligence in medicine," *Metabolism, Clinical and Experimental*, vol. 69, pp. S36–S40, Apr. 2017, https://doi.org/10.1016/j.metabol.2017.01.011.
47. A. Bhardwaj, S. Kishore, and D. K. Pandey, "Artificial intelligence in biological sciences," *Life*, vol. 12, no. 9, p. 1430, Sep. 2022, https://doi.org/10.3390/life12091430.
48. D. K. Saini, D. Yadav, S. Pabbi, D. Chhabra, and P. Shukla, "Phycobiliproteins from Anabaena variabilis CCC421 and its production enhancement strategies using combinatory evolutionary algorithm approach," *Bioresource Technology*, vol. 309, p. 123347, Aug. 2020, https://doi.org/10.1016/j.biortech.2020.123347.
49. R. H. Jhaveri, A. Revathi, K. Ramana, R. Raut, and R. K. Dhanaraj, "A review on machine learning strategies for real-world engineering applications," *Journal of Mobile Information Systems*, vol. 2022, pp. 1–26, Aug. 2022, https://doi.org/10.1155/2022/1833507.
50. N. E. Siedhoff, U. Schwaneberg, and M. D. Davari, "Machine learning-assisted enzyme engineering," in *Methods in Enzymology on CD-ROM/Methods in Enzymology*, Elsevier, 2020, pp. 281–315. https://doi.org/10.1016/bs.mie.2020.05.005.
51. C. Gu, G. B. Kim, W. J. Kim, H. U. Kim, and S. Y. Lee, "Current status and applications of genome-scale metabolic models," *Genome Biology*, vol. 20, no. 1, Jun. 2019, https://doi.org/10.1186/s13059-019-1730-3.
52. S. Srinivasan, W. R. Cluett, and R. Mahadevan, "Constructing kinetic models of metabolism at genome-scales: A review," *Biotechnology Journal*, vol. 10, no. 9, pp. 1345–1359, Sep. 2015, https://doi.org/10.1002/biot.201400522.
53. J. Almquist, M. Cvijovic, V. Hatzimanikatis, J. Nielsen, and M. Jirstrand, "Kinetic models in industrial biotechnology – Improving cell factory performance," *Metabolic Engineering*, vol. 24, pp. 38–60, Jul. 2014, https://doi.org/10.1016/j.ymben.2014.03.007.
54. L. Heirendt et al., "Creation and analysis of biochemical constraint-based models using the COBRA Toolbox v.3.0," *Nature Protocols*, vol. 14, no. 3, pp. 639–702, Feb. 2019, https://doi.org/10.1038/s41596-018-0098-2.
55. T. Oyetunde, F. S. Bao, J.-W. Chen, H. G. Martin, and Y. J. Tang, "Leveraging knowledge engineering and machine learning for microbial bio-manufacturing," *Biotechnology Advances*, vol. 36, no. 4, pp. 1308–1315, Jul. 2018, https://doi.org/10.1016/j.biotechadv.2018.04.008.
56. C. Angermueller, T. Pärnamaa, L. Parts, and O. Stegle, "Deep learning for computational biology," *Molecular Systems Biology*, vol. 12, no. 7, Jul. 2016, https://doi.org/10.15252/msb.20156651.
57. T. Oyetunde, D. Liu, H. G. Martin, and Y. J. Tang, "Machine learning framework for assessment of microbial factory performance," *PLoS One*, vol. 14, no. 1, p. e0210558, Jan. 2019, https://doi.org/10.1371/journal.pone.0210558.

58. L. Faure, B. Mollet, W. Liebermeister, and J.-L. Faulon, "A neural-mechanistic hybrid approach improving the predictive power of genome-scale metabolic models," *Nature Communications*, vol. 14, no. 1, Aug. 2023, https://doi.org/10.1038/s41467-023-40380-0.
59. S. Choudhury, M. Moret, P. Salvy, D. Weilandt, V. Hatzimanikatis, and L. Miskovic, "Reconstructing kinetic models for dynamical studies of metabolism using generative adversarial networks," *Nature Machine Intelligence*, vol. 4, no. 8, pp. 710–719, Aug. 2022, https://doi.org/10.1038/s42256-022-00519-y.
60. G. Dhanarajan, M. Mandal, and R. Sen, "A combined artificial neural network modeling–particle swarm optimization strategy for improved production of marine bacterial lipopeptide from food waste," *Biochemical Engineering Journal*, vol. 84, pp. 59–65, Mar. 2014, https://doi.org/10.1016/j.bej.2014.01.002.
61. S. M. J. Pappu and S. N. Gummadi, "Artificial neural network and regression coupled genetic algorithm to optimize parameters for enhanced xylitol production by Debaryomyces nepalensis in bioreactor," *Biochemical Engineering Journal*, vol. 120, pp. 136–145, Apr. 2017, https://doi.org/10.1016/j.bej.2017.01.010.
62. D. K. Saini et al "A multi-objective hybrid machine learning approach-based optimization for enhanced biomass and bioactive phycobiliproteins production in Nostoc sp. CCC-403," *Bioresource Technology*, vol. 329, p. 124908, Jun. 2021, https://doi.org/10.1016/j.biortech.2021.124908.
63. R. Czinkóczky, J. Sakiyo, E. Eszterbauer, and Á. Németh, "Prediction of surfactin fermentation with Bacillus subtilis DSM10 by response surface methodology optimized artificial neural network," *Cell Biochemistry and Function*, vol. 41, no. 2, pp. 234–242, Jan. 2023, https://doi.org/10.1002/cbf.3776.
64. E. G. F. Núñez et al., "Artificial intelligence approach for high level production of amylase using Rhizopus microsporusvar. oligosporusand different agro-industrial wastes," *Journal of Chemical Technology and Biotechnology*, vol. 92, no. 3, pp. 684–692, Aug. 2016, https://doi.org/10.1002/jctb.5054.
65. Z. Ahmad, D. Crowley, N. Marina, and S. Kr. Jha, "Estimation of biosurfactant yield produced by Klebseilla sp. FKOD36 bacteria using artificial neural network approach," *Measurement*, vol. 81, pp. 163–173, Mar. 2016, https://doi.org/10.1016/j.measurement.2015.12.019.
66. R. Jia, Y.-C. Song, D.-M. Piao, K. Kim, C.-Y. Lee, and J. Park, "Exploration of deep learning models for real-time monitoring of state and performance of anaerobic digestion with online sensors," *Bioresource Technology*, vol. 363, p. 127908, Nov. 2022, https://doi.org/10.1016/j.biortech.2022.127908.
67. R. Takors, "Scale-up of microbial processes: Impacts, tools and open questions," *Journal of Biotechnology*, vol. 160, no. 1–2, pp. 3–9, Jul. 2012, https://doi.org/10.1016/j.jbiotec.2011.12.010.
68. J. Smiatek et al., "Generic and specific recurrent neural network models: Applications for large and small scale biopharmaceutical upstream processes," *Biotechnology Reports*, vol. 31, p. e00640, Sep. 2021, https://doi.org/10.1016/j.btre.2021.e00640.
69. R. Lin, J. Chen, L. Xie, H. Su, and B. Huang, "Facilitating reinforcement learning for process control using transfer learning: Perspectives," *arXiv (Cornell University)*, Mar. 2024, https://doi.org/10.48550/arxiv.2404.00247.
70. Xu Pei, Lou Qun. "Solid state fermentation control method based on neural network and particle swarm algorithm," *Patent*, 07 May 2014.

71. T. R. Bjerk et al. "Biosurfactants: Properties and applications in drug delivery, biotechnology and ecotoxicology," *Bioengineering*, vol. 8, no. 8, p. 115, Aug. 2021, https://doi.org/10.3390/bioengineering8080115.
72. K. H. Nam, "AI-based protein models enhance the accuracy of experimentally determined protein crystal structures," *Frontiers in Molecular Biosciences*, vol. 10, Jun. 2023, https://doi.org/10.3389/fmolb.2023.1208810.
73. A. B. Sharon, E. F. Ahuekwe, E. G. Nzubechi, O. Oziegbe, and M. Oniha, "Statistical optimization strategies on waste substrates for solving high-cost challenges in biosurfactants production: a review," *IOP Conference Series Earth and Environmental Science*, vol. 1197, no. 1, p. 012004, Jun. 2023, https://doi.org/10.1088/1755-1315/1197/1/012004.
74. S. Nalini, D. Inbakandan, T. S. Dhas, and S. Sathiyamurthi, "Optimization of biosurfactant production by marine Streptomyces youssoufiensis SNSAA03: A comparative study of RSM and ANN approach," *Results in Chemistry*, vol. 3, p. 100223, Jan. 2021, https://doi.org/10.1016/j.rechem.2021.100223.
75. Z. Hussain, M. Babe, S. Saravanan, G. Srimathy, H. Roopa, and S. Boopathi, "Optimizing biomass-to-biofuel conversion," in *Circular Economy Implementation for Sustainability in the Built Environment*, Advances in Finance, Accounting, and Economics Book Series, IGI Global, 2023, pp. 191–214. https://doi.org/10.4018/978-1-6684-8238-4.ch009.
76. T. S. T. Ibrahim and T. S. Workneh, "The environmental impact of the sugar industry waste in Sudan," *Environmental Monitoring and Assessment*, vol. 195, no. 7, Jun. 2023, https://doi.org/10.1007/s10661-023-11401-8.
77. C. Samanvitha, H. Dharmendra, N. Nethravathi, and P. R. Velmurugan, "Artificial intelligence for sustainable futures," in *Advances in Computational Intelligence and Robotics Book Series*, IGI Global, 2024, pp. 129–154. https://doi.org/10.4018/979-8-3693-5380-6.ch006.

Chapter 10

Convergence of machine learning, deep learning, and IoT for plant disease detection

Akhilesh Kumar Srivastava, Chandrahas Mishra, Tribhuvan Mishra, Vaishnavi Mishra, and Saransh Chauhan

10.1 INTRODUCTION

Crops are the main source of food for all the humans of this planet; millions of tons of food are consumed regularly in different forms. In spite of this, according to reports, 828 million people are still starving for food in the world. Thus, producing ample amounts of food has become one of the necessities for all the nations. Because of pest control, various pesticides and insecticides are sprayed upon the crops, which not only kill the pests but also deteriorate the nutrients of the crops, which leads to various diseases in humans and also causes harm to the environment. Figure 10.1 shows how pesticides are sprayed over plants. They also cause water pollution, whenever they get into the ponds and lakes by rain water. The quality of food is also becoming an emerging concern for people. Pesticides and insecticides reach the ground through the roots of the plants and degrade the quality of soil in the long term.

Not just sowing crops but growing, storing, and transporting them are also big tasks. According to the Food and Agriculture Organization (FAO) of the United Nations, at least 40% of total global agricultural crops are lost due to pests. So, there is an ultimate need for regular monitoring of crops so that pests can be detected and classified at the earliest possible stage, thus causing the least damage to crops and a comparatively smaller amount of fertilizers and pesticides is used to protect the growth of that crop [2]. From Figure 10.2, we can observe that the use of chemicals on the crops at the global level as well as at the national level is very much alarming. Numerous studies also estimated that the overall world population will increase by almost 60%–70% by 2050. So, the world must be ready to face the increasing food demand in the future. Thus, there is an alarming need to detect and control the pests in the crop at the earliest possible stage.

Figure 10.3 gives a clear warning that the world must be ready for the upcoming hunger crisis if the right measures are not taken at the right time. With the help of technology and proper measures, the upcoming and ongoing problems can be tackled to a much greater extent. This global hunger might become the reason for some big crises around the globe.

 DOI: 10.1201/9781003494782-10

Various research and studies have taken place in the field of the above-mentioned subject. Pests' detection using automated technology is the main aim of the research studies done, and it is found that traps are the most common device that are being used for detecting pests in large farms. But the drawback is traps are not fully automated, that is, they are placed and collected at distinct positions by humans in the farms [5]. Various sensing technologies like thermal image sensors, Internet of Things (IoT) techniques for data collection, multispectral sensors, Red-Green-Blue sensors, and X-ray sensors are also used to detect and classify the pests. Image

Figure 10.1 Spraying of pesticides on the crops [1].

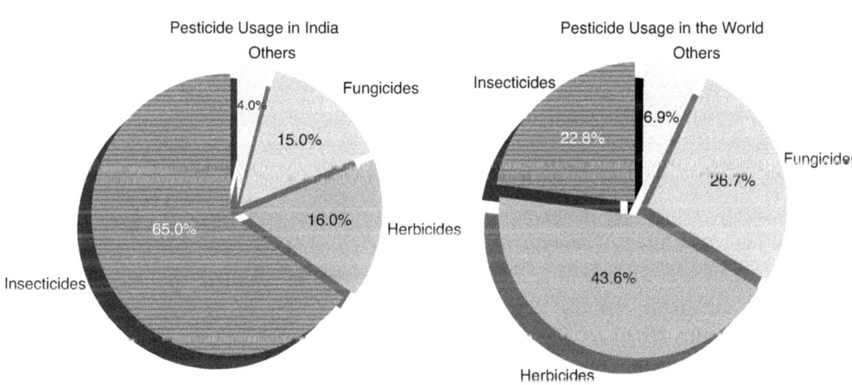

Figure 10.2 Assessment of toxicity of pesticides [3].

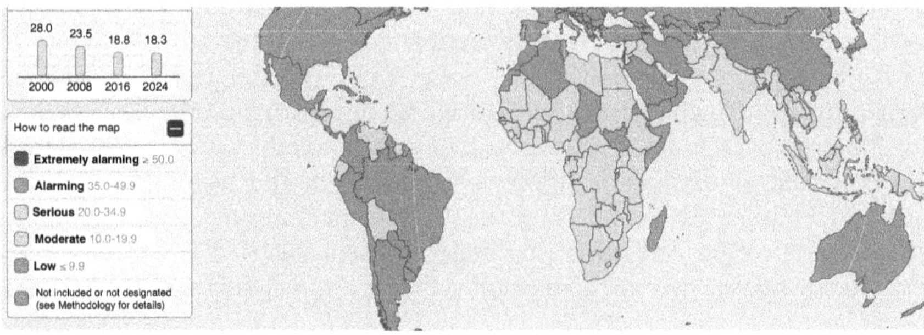

Figure 10.3 Graph depicting global hunger [4].

processing by using machine learning and deep learning algorithms has made the main contribution in detecting pests to date [6], [7].

10.2 EFFECTS OF PESTS IN CROPS

There are various problems which have been faced by the 7 billion population of the world in today's date; they may include climate change, global warming, various health diseases, water pollution, hunger, poverty, and many more, but one of the most alarming problems is the rapid increase in the need for food production because of the increase in population worldwide. A major challenge in food production is not only providing an ample amount of food, but also ensuring that the quality of food is good. Thus, food or crop protection from pests and insects has become an issue of concern. Various kinds of pests affect the crops in diverse ways, for example, pests like caterpillars, which are also called stem borers, affect the stem of the plants through their Larva [8], [9]. The most affected crops by the caterpillars include maize, rice, jackfruit, and mango. Thrips attack the buds and flowers of the plants, causing the leaves and flowers to crumble. They affect crops like groundnut, grapes, cotton, and rose. Figure 10.4 shows how insects feed on the leaves of plants and destroy their capabilities to perform photosynthesis, because of which the plant dies.

Pests not only affect the net income of farmers but also reduce the supply of food and the economic ratio of developing countries. India, being an agriculture-based developing nation, with such a large population, faces a lot of difficulty in managing the quantity and quality of food [10]. Qualitative food production has also become an important concern, as it plays a significant role in the economy of the country. Apart from the economic concern, the main and severe impact of pests and weeds is on the health and nutritional content of the crops. Due to pests, organic farming is being competitively converted into inorganic farming, which is causing various health hazards in day-to-day life [11]. So, detecting and controlling pests in the crops at an early stage is an alarming thought, which we all need to think of.

Figure 10.5 shows the main factors responsible for the loss of crops. Biotic and abiotic factors are mainly responsible for the damage to crops. Here, we will discuss the biotic factors responsible for crop losses. Every year, approximately 220 billion dollars (about $680 per person in the US) is spent on plant diseases. Apart from that, as the number of insects increases in crops, the need for pesticides is going up, which in turn reduces the quality of food [13].

Not only are the crops growing in fields being damaged by pests, but also the grains that are stored are not safe, which results in a threat to food security across the globe. Some bacteria, apart from damaging the crops, have a very bad impact on living beings around (human beings, animals, etc.) [15]. It was observed that for some crops, the losses due to pests were not curbed in the post-green revolution period, even after using chemicals and synthetic pesticides. The extent of synthetic chemicals for pests increased to a large

Figure 10.4 Plant infected by pests [12].

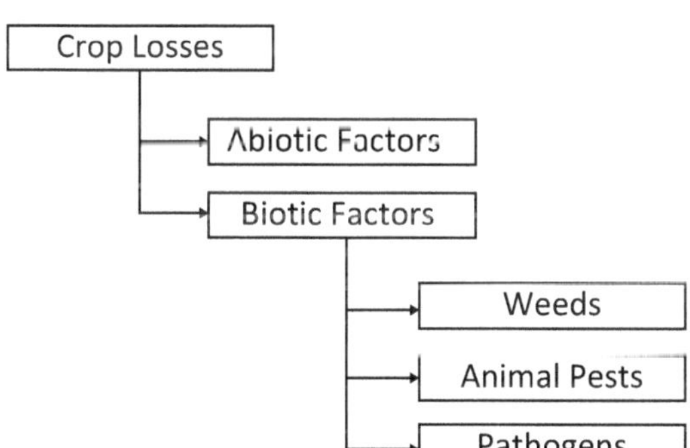

Figure 10.5 Flowchart depicting reasons for losses to crops [14].

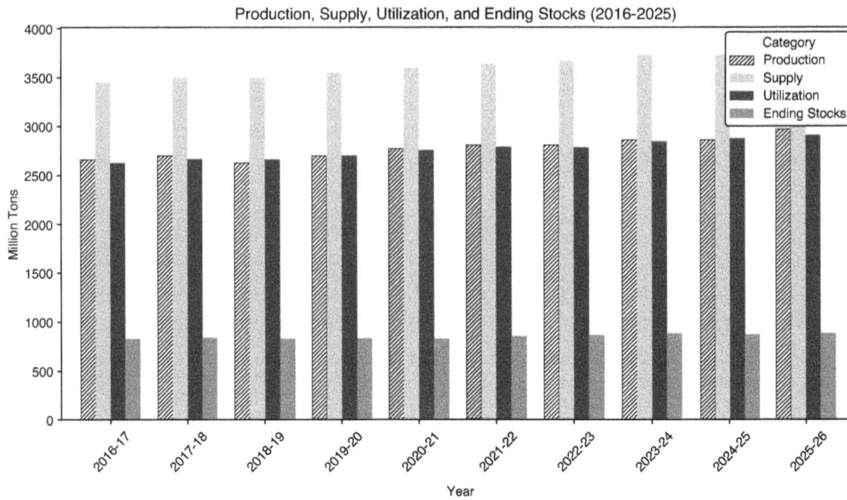

Figure 10.6 Graph showing cereal production, utilization, and stocks (in million tons) [17].

extent, while it didn't help in curbing the pests to that extent, although it helped in increasing the annual crop yield [16]. These losses also vary crop to crop, like for soyabean, the actual loss due to pests is estimated to be approximately 26%–29.1%, while for wheat, cotton, and maize, it is 31.1%, 37.01%, and 40%, respectively. The maximum loss is produced by weeds (34%), followed by animal pests and pathogens with a loss of 18% and 16%, respectively. As the population is increasing, the demand for more and more food and grains is increasing [14]. If we fail to protect our crops, it will lead to deaths due to hunger. Hence, it leads to loss of human life.

As per FAO, around 46 countries, including 33 African, 10 Asian, two Latin American and the Caribbean, and 1 in Europe, which now includes Ukraine and Sri Lanka as well, require external assistance for food [18]. From Figure 10.6 shown above, it can be clearly observed that the utilization of cereal production is increasing rapidly, while the production rate is comparatively slower, and stocks are also not increasing significantly with the demand. Humans have, till now, not found effective ways to fight abiotic factors causing losses, so it becomes essential to control biotic losses, in order to meet the current demands.

10.3 IDEA OF EARLY-STAGE PEST DETECTION AND ITS BENEFITS

Pests affect crops in various ways: they reduce the rate of photosynthesis (bacteria and fungi); some have the light-stealing effect on crops (some pathogens and weeds), while others affect the productivity of crops.

The main objective of early-stage pest detection is to detect the pest at its earliest possible stage, so that pests do not spread in the whole field and they can be controlled by using the least amount of pesticide that is taken in the targeted area only. This process will reduce the expenses of pesticides in the growing of the crop, which would indirectly reduce the soil and water pollution [19]. Early-stage pest detection is not only limited to reducing economic losses and improving environmental issues but also plays a significant role in reducing the total pesticide intake in the human body and thus improving human life. In the present time, there is an urgent need to look for possible solutions to this problem; otherwise, food scarcity will leave millions of people to die of hunger across the globe. Statistics of the years 1993–2020 shown in Figure 10.7 clearly depicts the increase in demand for cereals, meats, roots, and tubers in developing countries, developed countries, and of whole world. Thus, it is very much needed to detect the pests in the crops at a very early stage, which would help in improving the production of crops in sufficient amounts such that production fulfills the requirements of the food supply chain and thus improves the economy of the developing nations [4], [14].

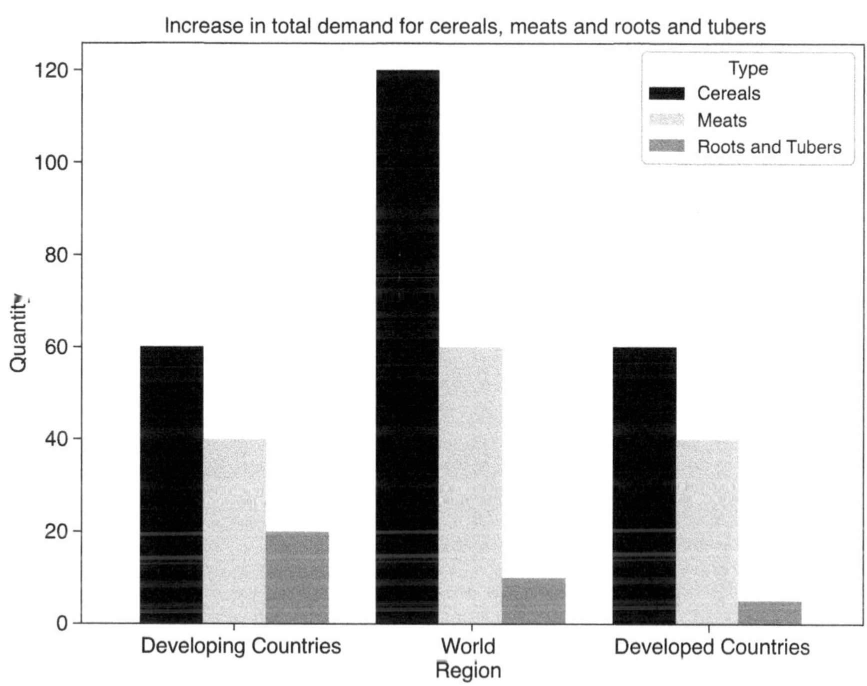

Figure 10.7 Bar graph showing increase in demand for cereals, meats, roots, and tubers, 1993–2020 [20].

10.4 USES OF VARIOUS TECHNOLOGIES IN DETECTING INSECTS IN CROPS

Mostly the farmers detect the presence of insects in crops by visual inspection; this method is less reliable, as when they come to know about a pest attack, it is extremely late. To make that process faster and reduce the losses, now image processing and machine learning technologies are applied in early-stage detection of pests. There is a lot of talk around IoT hardware devices for this purpose [21], [22]. Some of the previously applied techniques are discussed below:

Acoustic sensors are sensors that work on the principle of transmitting signals and listening to reflected signals to generate the output. Srivastav et al. [24] report that most of the acoustic sensors work on ultrasonic frequencies. This is the main reason that acoustic sensors may not be considered a reliable solution because with high speed of air, frequencies of the ultrasonic signal may change, which may result in changing the whole outcome. Figure 10.8 represents an acoustic sensor that transmits an acoustic signal and listens to the reflected sound. In an acoustic sensor, the signal is produced by moving the diaphragm in a back-and-forth motion [2].

Figure 10.9a shows an echo-acoustic trap and sensor; it gets activated when it senses some noise around itself in the surroundings and traps the insects whenever it enters inside the device. Figure 10.9b shows device equipped with seismic sensors, which converts vibrations due to seismic waves into electrical signals and when the insect comes near the seismic sensor, it records the disturbance and converts it into an electrical signal.

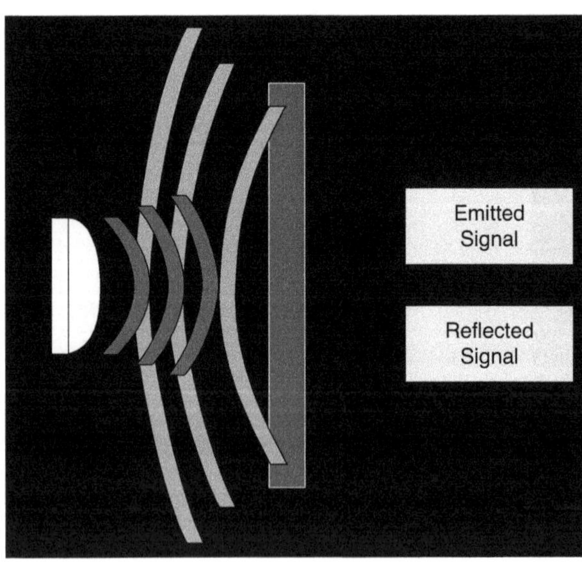

Figure 10.8 Design of an acoustic sensor [23].

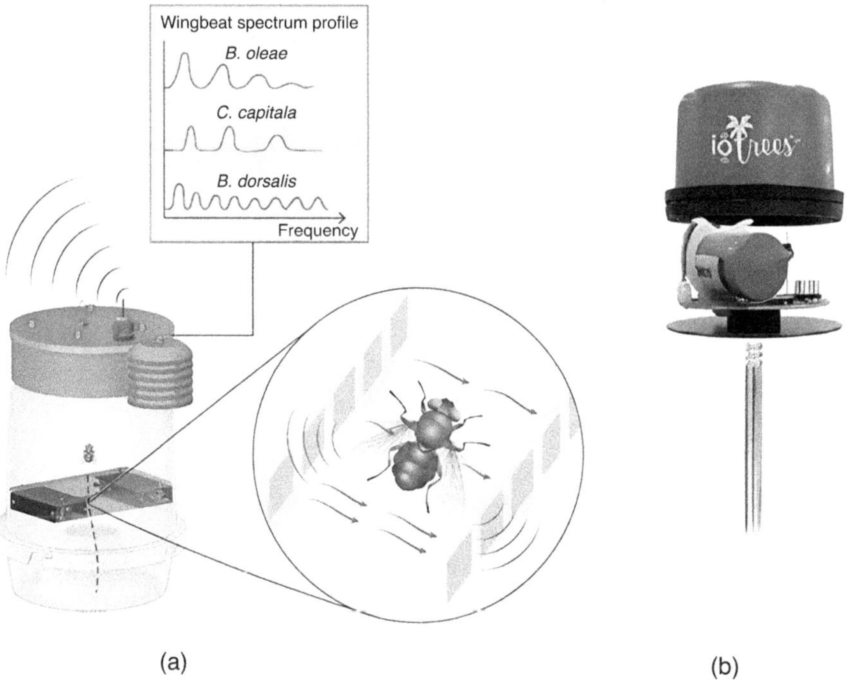

Figure 10.9 Working of an echo-acoustic trap sensor. (a) Optical Sensor based module for wingbeat frequency analysis (b) Insect monitoring device [25].

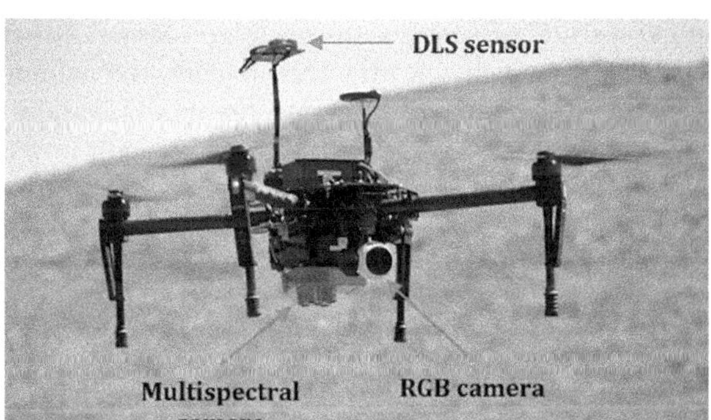

Figure 10.10 Drone capturing images of crops by a fitted camera [14].

The device shown in Figure 10.10 is a drone with a camera fitted that is being used to capture pictures of crops, and these pictures are analyzed by various machine learning and deep learning techniques to detect the presence of pests. This type of drone contains multispectral camera and RGB

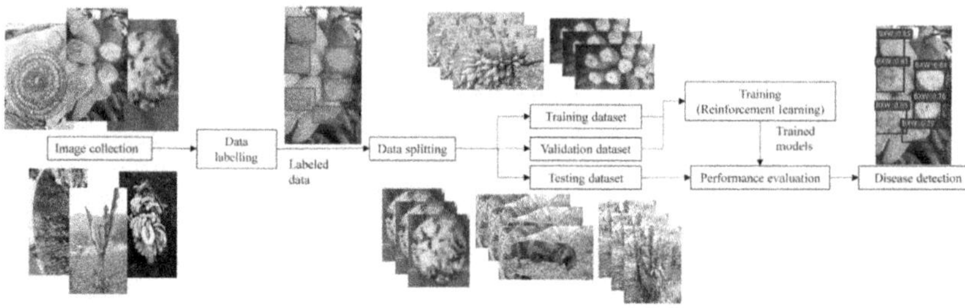

Figure 10.11 Deep learning model making a prediction about pests in images [3].

camera, which take images of the plants, and on the basis of reflectance of the plant in the infrared region and RGB region, it is decided that the plant is healthy or weak [7].

Deep learning is also emerging as a powerful technique in pest detection. Figure 10.11 shows a deep learning model predicting the presence of various pests feeding on a plant leaf. Nowadays, machine learning and deep learning techniques are widely used for pest detection. Kundur et al. [26] report that the whole process of deep learning is a combination of various steps like image collection, data labeling, and data splitting, which further include training the dataset, validating the dataset, and testing the dataset, model training, performance evaluation, and disease detection. The above image shows a full working IoT system. Initially, the raw images are collected from various IoT hardware devices like drone cameras and camera sensors installed in the field [27]. Then we need to feed the images to our machine learning model, as this is the task of supervised learning. Data labeling is an exhaustive task and requires huge manual efforts. After the data are labeled, they are split into the training set, the validation set, and the test set. The training set is used to train the model, validation data are used to validate the predictions, and testing data are kept unseen, and after validating them several times, the final model is used for testing data. After splitting the dataset, it is fed to a model for training, and the performance is evaluated. After performance evaluation, the disease is detected.

10.5 LITERATURE REVIEW

Srivastavet al. [24] in their research, discussed the use of acoustic sensors to solve the issue of pests in sugarcane. India is the world's second largest producer of sugarcane. Because of pests, a major amount of crop gets wasted; sugarcane is infected by 288 insects, of which 24 types of insects cause major damage. The reason for choosing acoustic sensors over other methods is that it is a self-monitoring system, frequency remote access, long battery life/low power system, and no need for specialists in the field.

They proposed a pest monitor and control system involving the use of acoustic sensors, which will monitor the noise level produced by the insects, and if it crosses a threshold level, it will notify the farmer of the area where the influx is occurring. This will make farmers aware of the pests, and they will be able to identify the region and take the necessary action accordingly [24].

Srinivas et al., in their research, discussed Red Palm Weevil (RPW, Rhynchophorus ferrigineus), which falls in the category of the most damaging pests in the coconut plant (Cocos nucifera). Because of its hidden nature of consuming food, RPW infestation is detected when the crops have already suffered severe damage. The acoustic action of RPW larvae (inside the base of leaves, an offshoot) consists of chewing the plant body, crawling over the plant, and emitting sounds. The sounds of larvae are captured by wireless networks of sensors fastened to palms and transmitted by access points to the server, covering the number of palms organized hexagonally for processing using MATLAB tools. The author team claimed that the method used is cheaper as compared to other available techniques for the detection of RPW larvae [7].

Pinhas et al., in their research, on automation acoustic detection of the presence of RPW (i.e., Red Palm Weevil) highlighted the use of sound-based activity of RPW in shoots and implemented it in a framework setup. RPW is a main pest of horticulture and ornamental palm species in the Middle East, Asia, and the Mediterranean region. The RPW larvae make the palm dead by boring deep in the trunks, crowns, and offshoots. The methodology applied was utilizing VQ (Vector quantization) or Gaussian mixture modeling. The author team claimed that with their efforts, they achieved a detection ratio of approximately 98.9%. Their study depicted that it is possible to detect RPW sounds with the use of the mathematical method of speech recognition and commercial recording devices [27].

Betty Martin et al., in their research on the acoustic activity of RPW, noted that RPW is the most dangerous pest feeding on coconut crops; the author team measured the frequencies of sounds produced by RPW when they feed on coconut trees. The author team measured the sound by using a digital voice recorder with a sensor microphone; they measured the sound of movement of RPW in trees and divided it into three categories, which are: (1) Normal movement, (2) Slow movement, and (3) lengthy movement; they then assigned a range of frequencies according to the type of movement they make. Similarly, they measured the sound of their eating and divided it into two categories: (1) Clear eating and (2) slow eating; they then assigned these categories a frequency according to the type. This could help to detect the movement of RPW by detecting its unique traits like biting, eating, and movement [28]. They collected the analog signals and converted them into digital signals whenever they were required. The pest sound can be detected by observing the spikes as sounds of insects [10].

Betty Martin and Vimala Juliet, in their research based on differentiating pest sound from human sounds, aimed to differentiate between human sound and pest sound so that the wrong elucidation would not confirm the sound-based activities of the pest. They collected the data of the sounds of pests and humans who installed sensors for recording the voice from the Pattukottai district of Tamil Nadu, India. After the collection of datasets, they led it through Mel frequency coefficient characteristics (MFCC) feature extraction, through which they arrived at the conclusion that human whispers can be successfully discriminated from pest sounds by using MFCC and VQ. They formed a codebook, which is a matrix, by using Lloyd's algorithm. The codebook can be created by either selecting the code vectors randomly or by calculating the centroids that will form the codebook. The author team created the codebook by calculating the centroids. MFCC are applied to the input for the identification of sound, and VQ is a quantization method used for dealing with vectors. The error calculation was done by estimating the Euclidean distance between each code vector and each training vector. At the end, they successfully discriminated between the human whispers and pest sounds. The author team also suggested carrying out this work using LPC instead of MFCC [29].

Bhadane et al. focused their research on pests' detection at the earliest stage using image processing and interpretation. The author's team started their work with the thought of detecting the pest at the most initial stage, that is, eggs and larvae, but because of their critical dimensions, it turned out to be a challenging task, and thus they performed their research firstly on the adults (whiteflies) [30]. Techniques like advanced and cognitive computer vision, image processing, scanner image acquisition, sampling optimization, and pattern recognition are used to get high resolution image of the leaf, so that symptoms of pest attacks can be detected and the model results as an automated monitoring system. In the paper, the entire process is divided into various stages, and the first among them is object extraction [31]. Object extraction is the technique of extracting the important segments from the image; it is done with the help of the difference in the color of the background and foreground regions. Another technique, which is used in object extraction, is edge detection; it is a type of segmentation technique through which various parameters of an object such as area, perimeter, etc. can be extracted. Edge detection is done by identifying sharp discontinuities in the image. These processes are carried out by using various operators and algorithms like the Soble operator is used in the edge detection algorithm. After the object is extracted, then the feature extraction is done, which includes calculating the infected area and calculating the number of pests. This research claims that their prototype is an automated and robust monitoring system which can cover three times larger area and can detect whiteflies three times faster [10].

Wen et al. focused their research on moth identification with the help of the poses of moth. Researchers' main aim was to identify moths from the field images having a cluttered background. They took samples of moths using sticky traps, which were distributed over the field. For the identification process, a total of 9 species and 480 samples were collected, which also include multiple poses of moths and top and side views of 248 male moths. The entire process of moth identification is subdivided into various stages, which include image segmentation, feature extraction, and image identification. Firstly, segmentation was performed against a chaotic background having various impurities and uneven light distribution, which include grid lines of moths and broken body parts of insects, which helped to distinguish the moth from the image of the field. After that, the proficient feature had been extracted for the input of the deep learning model to classify the pose of the moth. The proposed model (IpSDAE) of deep learning in the research is the improved version of stacked denoising auto-encoders (SDAE). The research claims that this method of moth identification through pose attained an accuracy of 98.1% over 428 fields and 480 labs. Also, the pose identification method enhanced moth identification through other techniques like Linear Regression Classifier (LRC), Radial Basis Function (RBF) Network, and so forth [32].

Kang et al. conducted research on the identification of butterflies through posture using an artificial neural network. This study clarifies that automation of the described task can be done in two stages: (1) preprocessing and (2) feature extraction; in the research, a single neural network with a multi-class pattern classifier is used to identify 15 different breeds of butterflies. In this study, the BLS entropy is used as an input to feed into the artificial neural network so that butterfly specimens can be identified when examined from various angles. After the input feeding process is done, all the image specimens are converted into binary images, and then 180 pixels are evenly distributed on the body and the BLS entropy at each point is calculated. After calculation, it is found that in various cases, effective results are found, whereas in some of the cases, results are not that efficient. The study clarifies that computational proficiency of the identification process can be increased by using the results of comparison of BLS entropy at different uniformly and non-uniformly spaced points. The study claims that the suggested model can be implemented in embedded systems like Android and OS in mobile phones, as it uses a simple algorithm and mathematical calculations [33], [34], [35], [36].

Tripathi et al. researched the detection and classification of plant diseases in the leaves of the plants [37]. The researcher's team considered plant leaves, which were affected by early blight, late blight, powdery mildew, and Septoria and applied their proposed model over them, which includes image processing, neural network, and color co-occurrence technique to extract the desired results. In the proposed model, firstly, a device structure was created, which was independent of color transformation and changed the color of the image to the space color specified in the structure. After that, with the help of k-means clustering, the whole image of the leaf is divided into four clusters, in which one or more clusters are infected by the disease. In the research, clusters were used for feature extraction, and the extracted features were trained in a way so that they can be used as input for training a neural network and thus detecting the leaf image. A MATLAB software program is written in the .mat format; this MATLAB routine loads all the data files and modifies the data files according to the proposed structure. After that, the neural network is fed using extracted features as input in forward–backward manner and thus the subsequent leaf disease is detected. The researcher team suggested that the recommended model is a simple automated model, which is efficient enough to detect plant leaf disease and can be developed for future use [37].

Venugoban et al. researched the classification of paddy field insect pests using Gradient-based feature classification. The main advantage of the suggested technique is that it can detect many reliable features, and it is invariant to changes in the pose and scale. A sample of 20 paddy field insect pests was taken in a system, and the entire system was compared and analyzed with SURF and SIFT descriptors using HOG features and bag-of-words approach. In this study, every feature of the image is quantified to form clusters. The HOG+SURF framework is used in the study, as it contains several advantages, which include capturing of edge and gradient structure, invariant to rotational changes and local geometry. The PCA technique is used to reduce the dimensions of HOG features. Inclusively, the model and technique presented in the paper is efficient enough to draw the desired results because of its simplified implementation. The study claims that the model had achieved 90% accuracy in detecting the image of insects in a paddy field. The author's team also suggested that in the future, the proposed model can be integrated with mobile phones after some modifications, which would make it easy for the farmers to detect the disease or the insect at the right time with less labor [8].

10.5.1 Drawback of image sensing technology

- Reviewing several research papers has made it clear in our minds that all the image processing techniques can only work when there is something to capture, and in the captured image, there must be some feature to extract from the image, which clearly indicates that there must be some amount of distortion that had to be done in plants because of pests, and this distortion is detected by the proposed models, which we called automated systems [38].
- Even the high-resolution images are not efficient enough to extract features directly because of the cluttered background, uneven light distribution, and dust particles [39].
- It is a complicated task to take the image of pests because of the continuous movement of the pests. Two different images of the same pest from two viewpoints would be captured, extracting distinctive features from different images, and considering all the features is a bit challenging task.
- It is an arduous task to apply the model in the field; amplification of the hardware model in the field requires a lot of care during application.

10.5.2 Drawback of acoustic sensing technology

- A big downside of the classical methods is that they are only able to determine the higher pest densities (more than one insect in a kg of grain), which makes them systems of low sensitivity. The sounds also include other sounds from fields, and it becomes difficult to differentiate between those sounds and insect sounds.
- The unwanted noise of the surroundings interferes a lot and makes it difficult to get exact data [40].
- Hardware devices have limited accuracy and scope; they cannot record all the frequencies. A lot of information that can be used to make useful inferences remains uncaptured by devices [28].

Table 10.1 shows the recent works that have been done in order to detect various species of insect group flies. It also shows the sensors/devices used, along with the automation detection technique involved in the research [41], [20], [42].

Table 10.2 shows the recent works that have been done in order to detect various species of the insect group, white flies and snow flies; it also shows

Table 10.1 Detection of species of the insect group flies

Insect group	*Species of the insects*	*Sensors/devices used*	*Automation detection technique used*	*Authors of the paper*
Flies	Bactrocera dorsalis	Wind speed sensors, temperature, humidity sensors, an infrared counting device, GPS receiver, and GSM.	Remote monitoring platform-based system and the use of HCP (host check platform)	Jiang et al. [18]
Flies	Dacus dorsalis	Wind speed sensing device, temperature, humidity sensors, Infrared counting device, GPS receiver, GSM. Microcontroller (MSP430F449 chip), WSN.	Remote monitoring platform-based system and the use of HCP (host check platform)	Okuyama et al. [43]
Flies	Dacus dorsalis	GSM and ZigBee, with a remote sensing information gateway, a host control platform, and wireless monitoring nodes. The WMNs send the obtained data to the RSIG, and the RSIG delivers the data to the database server for analysis.	Cloud services, Image detection	Liao et al. [44]
Flies	Olive fruitfly	Webcams connected to a McPhail trap	OTSU algorithm threshold, Gauss blur filter	Doitsidis et al. [3]

Table 10.2 Detection of species of insect group white flies and snow flies

Insect group	*Species of the insects*	*Sensors/devices used*	*Automation detection technique used*	*Authors of the paper*
Whiteflies	Not specified	Image dataset	Classifiers like ANN, SVM, Bayes classifier, k-nearest neighbor classifier. Gray-level co-occurrence matrix (GLCM).	Dey et al. [45]
Whiteflies	Not specified	Image dataset	Structure, texture, and gray-scale strength analysis.	Ghods and Shojaeddini [46]
Snow flies	Bemisia tabaci	Yellow sticky net	Grayscale, boundary tracking, counting, binarization, and image transformation	Qiao et al. [47]
Snow flies	Bemisia tabaci	Digital image capturing device, plants, and adhesive traps.	Machine learning and eccentricity area features	Soliz-Sánchez et al. [37]

Table 10.3 Detection of various species of the insect group Lepidoptera

Insect group	*Species of the insects*	*Sensors/devices used*	*Automation detection technique used*	*Authors of the paper*
Lepidoptera	Pyralididae (species unspecified)	Automated toy vehicle equipped with an image-capturing device	Changing into Hue, separation, value color space, taking out the H spatial matrix, Nobuyuki Otsu segmentation, and object contour detection on the basis of Hu activity	Liu et al. [45]
Lepidoptera	Pyralididae (species unspecified)	Automated toy vehicle equipped with an image-capturing device	Gaussian Mixture Model, Distance Regularization Level Set Evolution, and Aggregation Dispersion Variance	Zhao et al. [50]
Lepidopteran	Tortricidae, Argyrotaenia velutinana, etc	Webcam camera	Deep neural network generation together with color, structure, and numerical aspects.	Wen et al. [32]
Lepidopteran	Cydia pomonella	Commercial trap with modified comm. trap.	Visual remote inspection	Guarnieri et al. [51]

(Continued)

Table 10.3 (Continued) Detection of various species of the insect group Lepidoptera

Insect group	*Species of the insects*	*Sensors/devices used*	*Automation detection technique used*	*Authors of the paper*
Lepidopteran	Parantica sita, Dilipa fenestra, etc.	Images dataset	Artificial Neural Network and BLS entropy	Kang et al. [33]
Lepidoptera	Bicyclus anynana	Digi camera, Nikon SMZ1500dissecting microscope	Machine learning algorithms	Silveira and Monteiro [52]
Lepidoptera	50 species	Camera	GLCM of image pixels, Histogram of MSC (multi-scale curvature), and KNN classifier.	Li and Xiong [46]
Lepidopteran	19 species belonging to Papilinidae group	Images Dataset	Color and texture filter with GLCM (grey-level co-occurrence matrix)	Kaya et al[53]

the sensors/devices used, along with the automation detection technique involved in the research.

Table 10.3 shows the recent works that have been done in order to detect various species of the insect group Lepidoptera; it also shows the sensors/devices used, along with the automation detection technique involved in the research [48], [49].

Table 10.4 shows the recent work done to detect various species of the insect group palm tree weevil and the sensors/devices used, along with the automation detection technique involved in the research.

10.6 FUTURE SCOPE

There are plenty of research studies done in the field of agriculture, especially on early-stage pest detection [58], [11], [39], [56]. Various approaches, including artificial intelligence and machine learning techniques, image processing, audio processing, and thermal sensing, have been used, and accordingly, different results have been obtained. In spite of getting these results, it is difficult to detect pests at the earliest stage possible. Here is a summary of various approaches that the author team found useful: -

- Search for some methods based on the biological response of plants to the insects or pests feeding over them. For example, if we look at the human body, sometimes they get fever; the reason is that most fevers are caused by infections. They get fever, as their body is trying

Table 10.4 Detection of various species of the insect group palm tree weevil

Insect group	*Species of the insects*	*Sensors/devices used*	*Automation detection technique used*	*Authors of the paper*
Palm tree weevil insect	Rhynchophorus ferrugineus	AED-2000 and a digital recorder.	Digitize, Analyze, View, Insect Sounds: Conventionally written insect signal analysis program.	Pinhas et al. [27]
Palm tree weevil insect	Rhynchophorus ferrugineus	Piezo-electric sound transmitter with 50-mm diameter, model MP13. Raven Pro 64 was used for recordings.	Multivariate distribution, multivariate classifier	Hetzroni et al. [43]
Palm tree weevil insect	Rhynchophorus ferrugineus	Sound transmitter, radio amplifying device, transmitter, and repeater were connected to a wireless sensor network.	Threshold filter, low-pass filter, anti-aliasing image filter, and compression (down-sampling).	Srinivas et al. [7]
Palm Weevil	Rhynchophorus ferrugineus	A magnetic cartridge is used.	Time-frequency analysis and wavelet analysis.	Siriwardena et al. [54]
Palm Weevil	Rhynchophorus ferrugineus	POM-3542P-R acoustic Sound transmitter, 4th order passband, continuous-time filter (MAX274)	Analysis of frequency spectra.	Gutiérrez et al. [55]
Palm Weevil	Qrycites rhinoceros, Rhynchophorus ferrugineus.	Digital Recorder	Los Mel Frequency Cepstral Coefficient, Analysis of frequency and decibels, and Euclid's distance.	Martin et al. [56]
Palm Weevil	Rhynchophorus ferrugineus	Digital audio recording device.	Vector quantization and Mel frequency coefficient.	Martin and Juliet [57]

to eliminate the bacteria or viruses that cause any infection. Most of these viruses and bacteria do good when the body is at a normal temperature. But in fever, it becomes difficult for them to survive. Fever activates the immune system of our body. Similarly, plants also release toxic chemicals whenever any insect feeds on them. When tracking the concentration of these chemicals, if it rises over a certain threshold limit, then it can be said that the plant has been attacked by some pest; this will result in pest detection at an exceedingly early stage, because this is how the attack of insects starts over the plants.

- The scope of many studies is limited to images that are generated in the visible range, but the electromagnetic waves formed also have a lot of data and information, which can be used in an effective way to get more accurate results. Future research studies should have a method to fuse information from various sources and to carry out pest detection with the help of that information.
- Most of the research studies are currently carried out in a controlled environment. There are proper arrangements for lightning, and the entry of unwanted agents is not possible (like in fields, many a time, some bird or animal can harm the equipment), so the data collection and monitoring techniques should be optimized and designed to work in an open natural environment. Currently, it is also seen that there is a lot of human intervention required in these processes; this should be reduced by using unmanned aerial vehicles and IoT monitoring equipment.
- In early-stage pest detection using image processing, it is very difficult to detect the pests because the symptoms are not obvious, as it is the starting stage. But if pests could be detected at this stage, then the crops would be saved with a bare minimum of damage, but image processing alone is not sufficient to capture this type of sensitive information. Hence, there is a need to combine factors like temperature, humidity, and light around the area of which the image is to be captured, as a good image can only be captured in proper sunlight and clean weather .
- Currently most of the pest detection methods are deep learning-based, and that too in supervised learning [59]. Now, supervised learning requires labeled data, and the collection and maintenance of labeled data require a lot of manpower; so their applicability of unsupervised deep learning methods should be explored. Unsupervised deep learning methods will save a lot of manpower and efforts. Developing some new algorithms with human vision-like capabilities and the ability to use prior knowledge like the human brain is also a new way that can be explored .
- The sensors currently being used for collecting data in fields should be more tuned to capture frequencies of sound from insects and filter the unwanted noises; this will not only help in reducing the volume of data collected, but it will also reduce the time required for data preprocessing. Current deep learning techniques require a lot of computing

resources and very powerful systems with enhanced graphic cards to train the models, which makes them difficult to implement on mobile platforms with less resources. Hence, there is a need to search for less complex ways to make these models, so that deployment can be easier .
- Biorecognition elements like DNA/RNA, enzymes, and antibodies can also be used to identify pests and insects at an early stage, and it is also a more accurate way to capture correct information; it has very less chance of false signals.

10.7 CONCLUSION

This survey focuses on the most recent methods used for pest detection. This includes IoT, machine learning, and deep learning-based techniques for detecting, forecasting, and classification of pests.

The world's population is exploding at an alarming rate, and failing to match production will create huge problems. The goal of this paper is to review the recent research studies and provide some suggestions and other future scopes that can be worked upon to overcome these problems. After reviewing ample research papers involving various mathematical models and innovative technologies, the author team created a table that highlights the various techniques used for pest detection along with the equipment and technology used. After an in-depth study of various researches, the author team found that there are many studies being carried out, but they are being carried out in a controlled environment, and in these environments, good accuracy is obtained, but the real challenge comes when these techniques are deployed in an open environment.

In the present day, as deep learning algorithms and libraries such as PyTorch and TensorFlow are advancing, we are getting more and more accurate results. Methods of computer vision involving convolutional neural network models give more exact results when compared to traditional and old methods involving manual feature extraction [49]. Use of transfer learning is eliminating the biggest problem of deep learning, which is the requirement of a large dataset for training purposes. Transfer learning makes it possible to use a model that has already been trained millions of times on a wide variety of datasets.

In case of auditory datasets, it has been seen that audio files can be converted to images and can be analyzed to give more accurate results, which also made it possible to use audio data for detection. New sensors are being developed that remove a lot of noise from data while collecting it and have features of catching a particular frequency accurately will be very useful for pest detection using the noise of pests.

Although it is not a fact that all work has been done in this domain, there are still hundreds of challenges that need to be tackled like, when it comes to pest detection in open fields, there are still a lot of challenges. The author team has suggested some future work, which includes some techniques like biorecognition and biological response detection as well; these techniques are less explored, but as the study continues, they can provide much more accurate results.

REFERENCES

[1] J. Villalobos, S. Flores, P. Liedo, and E. A. Malo, "Mass trapping is as effective as ground bait sprays for the control of Anastrepha (Diptera: Tephritidae) fruit flies in mango orchards," *Pest Management Science*, vol. 73, no. 10, pp. 2105–2110, 2017.

[2] M. A. Ali, M. S. Sayeed & S. F. A. Razak, "Employing IoT and pest sound analysis with multi-feature and multi-deep learning networks for detecting, preventing and controlling the pests in expansive farmland," *International Journal of Machine Learning and Cybernetics*, vol. 16, pp. 7763–7792, 2025. Doi: 10.1007/s13042-025-02685-y

[3] J. Liu and X. Wang, "Plant diseases and pests detection based on deep learning: A review," *Plant Methods*, vol. 17, no. 1, Feb. 2021, doi: 10.1186/s13007-021-00722-9.

[4] T. Liu, W. Chen, W. Wu, C. Sun, W. Guo, and X. Zhu, "Detection of aphids in wheat fields using a computer vision technique," *Biosystems Engineering*, vol. 141, pp. 82–93, Dec. 2015, doi: 10.1016/j.biosystemseng.2015.11.005.

[5] L. Doitsidis, G. N. Fouskitakis, K. N. Varikou, I. I. Rigakis, S. A. Chatzichristofis, A. K. Papafilippaki, and A. E. Birouraki, "Remote monitoring of the Bactrocera oleae (Gmelin) (Diptera: Tephritidae) population using an automated McPhail trap," *Computers and Electronics in Agriculture*, vol. 137, pp. 69–78, Apr. 2017, doi: 10.1016/j.compag.2017.03.014.

[6] B. Prasath and M. Akila, "IoT-based pest detection and classification using deep features with enhanced deep learning strategies," *Engineering Applications of Artificial Intelligence*, vol. 121, p. 105985, 2023, doi: 10.1016/j.engappai.2023.105985.

[7] S. Srinivas, K. S. Harsha, A. Sujatha, and N. Kumar, "Efficient protection of palms from RPW Larvae using WSN," *IJCSI*, vol. 10, no. 3, pp. 192–200, May 2013.

[8] K. Venugoban and A. Ramanan, "Image classification of paddy field insect pests using gradient-based features," *International Journal of Machine Learning and Computing*, vol. 4, no. 1, pp. 1–5, Feb. 2014, doi: 10.7763/ijmlc.2014.v4.376.

[9] G. Bhadane, S. Sharma, and V. B. Nerkar, "Early pest identification in agricultural crops using image processing techniques," *International Journal of Electrical, Electronics and Computer Engineering*, vol. 2, no. 2, pp. 77–82, 2013.

[10] G. Bhadane, S. Sharma, and V. Nerkar, "Early pest identification in agricultural crops using image processing techniques," *International Journal of Electrical, Electronics and Computer Engineering*, vol. 2, no. 2, pp. 77–82, 17th Oct. 2013.

[11] D. Popescu, A. Dinca, L. Ichim, and N. Angelescu, "New trends in detection of harmful insects and pests in modern agriculture using artificial neural networks. A review," *Frontiers in Plant Science*, vol. 14, Nov. 2023, doi: 10.3389/fpls.2023.1268167.
[12] A. Gutierrez, A. Ansuategi, L. Susperregi, C. Tubío, I. Rankić, and L. Lenža, "A benchmarking of learning strategies for pest detection and identification on tomato plants for autonomous scouting robots using internal databases," *Journal of Sensors*, vol. 2019, pp. 1–15, May 2019, doi: 10.1155/2019/5219471.
[13] C. Xia, T.-S. Chon, Z. Ren, and J.-M. Lee, "Automatic identification and counting of small size pests in greenhouse conditions with low computational cost," *Ecological Informatics*, vol. 29, pp. 139–146, Sep. 2014, doi: 10.1016/j.ecoinf.2014.09.006.
[14] W. Li, P. Chen, B. Wang, and C. Xie, "Automatic localization and count of agricultural crop pests based on an improved deep learning pipeline," *Scientific Reports*, vol. 9, no. 1, May 2019, doi: 10.1038/s41598-019-43171-0.
[15] M. A. Ebrahimi, M. H. Khoshtaghaza, S. Minaei, and B. Jamshidi, "Vision-based pest detection based on SVM classification method," *Computers and Electronics in Agriculture*, vol. 137, pp. 52–58, Apr. 2017, doi: 10.1016/j.compag.2017.03.016.
[16] C.-Y. Lu, D. J. A. Rustia, and T.-T. Lin, "Generative adversarial network based image augmentation for insect pest classification enhancement," *IFAC-PapersOnLine*, vol. 52, no. 30, pp. 1–5, Jan. 2019, doi: 10.1016/j.ifacol.2019.12.406.
[17] D. J. A. Rustia, C. E. Lin, J.-Y. Chung, Y.-J. Zhuang, J.-C. Hsu, and T.-T. Lin, "Application of an image and environmental sensor network for automated greenhouse insect pest monitoring," *Journal of Asia-Pacific Entomology*, vol. 23, no. 1, pp. 17–28, Nov. 2019, doi: 10.1016/j.aspen.2019.11.006.
[18] K. R. Gavhale and U. Gawande, "An overview of the research on plant leaves disease detection using image processing techniques," *IOSR Journal of Computer Engineering*, vol. 16, no. 1, pp. 10–16, Jan. 2014, doi: 10.9790/0661-16151016
[19] S. Xuesong, L. Zi, S. Lei, W. Jiao, and Z. Yang, "APHID identification and counting based on smartphone and machine vision," *Journal of Sensors*, vol. 2017, pp. 1–7, Jan. 2017, doi: 10.1155/2017/3964376.
[20] M. P. Gautam, H. Singh, S. Kumar, V. Kumar, G. Singh, and S. N. Singh. "Diamondback moth, Plutella xylostella (Linnaeus)(Insecta: Lepidoptera: Plutellidae) a major insect of cabbage in India: A review." *Journal of Entomology and Zoology Studies*, vol. 6, no. 4, pp. 1394–1399, 2018.
[21] S. M. Pedersen and K. M. Lind, "Precision Agriculture – From Mapping to Site-Specific Application," in *Progress in Precision Agriculture*, Springer, 2017, pp. 1–20. doi: 10.1007/978-3-319-68715-5_1.
[22] J.-A. Jiang et al., "A GSM-based remote wireless automatic monitoring system for field information: A case study for ecological monitoring of the oriental fruit fly, Bactrocera dorsalis (Hendel)," *Computers and Electronics in Agriculture*, vol. 62, no. 2, pp. 243–259, Feb. 2008, doi: 10.1016/j.compag.2008.01.005.
[23] V. L. Nangai and B. Martin, "Interpreting the acoustic characteristics of RPW towards its detection: A review," in *IOP Conference Series: Materials Science and Engineering*, 2017, vol. 225, no. 1, p. 012178. IOP Publishing.
[24] N. Srivastav, G. Chopra, P. Jain, and B. Khatter, "Pest monitor and control system using WSN with special reference to acoustic device," *ICEEE*, Jan. 2013.

[25] I. Potamitis, I. Rigakis, N. Vidakis, M. Petousis, and M. Weber, "Affordable bimodal optical sensors to spread the use of automated insect monitoring," *Journal of Sensors*, vol. 2018, pp. 1–25, Jan. 2018, doi: 10.1155/2018/3949415.
[26] N. C. Kundur and P. B. Mallikarjuna, "Pest detection and recognition: An approach using deep learning techniques," *2022 Fourth International Conference on Cognitive Computing and Information Processing (CCIP)*, pp. 1–6, 2022, doi: 10.1109/CCIP57447.2022.10058692.
[27] J. Pinhas, V. Soroker, A. Hetzroni, A. Mizrach, M. Teicher, and J. Goldberger, "Automatic acoustic detection of the red palm weevil," *Computers and Electronics in Agriculture*, vol. 63, no. 2, pp. 131–139, Apr. 2008, doi: 10.1016/j.compag.2008.02.004.
[28] W. Li, T. Zheng, Z. Yang, M. Li, C. Sun, and X. Yang, "Classification and detection of insects from field images using deep learning for smart pest management: A systematic review," *Ecological Informatics*, vol. 66, p. 101460, Oct. 2021, doi: 10.1016/j.ecoinf.2021.101460.
[29] B. Martin and V. Juliet, "Discriminating human whispers from pest sound during recordings in coconut palm grooves using MFCC and vector quantization," *International Journal on Applied Bio-Engineering*, vol. 4, no. 2, pp. 29–33, Jan. 2010, doi: 10.18000/ijabeg.10067.
[30] L. O. Solis-Sánchez, J. J. García-Escalante, R. Castañeda-Miranda, I. Torres-Pacheco, and R. Guevara-González, "Machine vision algorithm for whiteflies (Bemisia tabaci Genn.) scouting under greenhouse environment," *Journal of Applied Entomology*, vol. 133, no. 7, pp. 546–552, Mar. 2009, doi: 10.1111/j.1439-0418.2009.01400.x.
[31] Y. Sun, H. Cheng, Q. Cheng, H. Zhou, M. Li, Y. Fan, G. Shan, L. Damerow, P. Schulze Lammers, and S. B. Jones, "A smart-vision algorithm for counting whiteflies and thrips on sticky traps using two-dimensional Fourier transform spectrum," *Biosystems Engineering*, vol. 153, pp. 82–88, Nov. 2016, doi: 10.1016/j.biosystemseng.2016.11.001.
[32] C. Wen, D. Wu, H. Hu, and W. Pan, "Pose estimation-dependent identification method for field moth images using deep learning architecture," *Biosystems Engineering*, vol. 136, pp. 117–128, Jun. 2015, doi: 10.1016/j.biosystemseng.2015.06.002.
[33] S.-H. Kang, J.-H. Cho, and S.-H. Lee, "Identification of butterfly based on their shapes when viewed from different angles using an artificial neural network," *Journal of Asia-Pacific Entomology*, vol. 17, no. 2, pp. 143–149, Jan. 2014, doi: 10.1016/j.aspen.2013.12.004.
[34] C. Doorenweerd, L. Leblanc, A. L. Norrbom, M. San Jose, and D. Rubinoff, "A global checklist of the 932 fruit fly species in the tribe Dacini (Diptera, Tephritidae)," *ZooKeys*, no. 730, p. 19, 2018.
[35] I. Potamitis, I. Rigakis, and N. A. Tatlas, "Automated surveillance of fruit flies," *Sensors*, vol. 17, no. 1, p. 110, 2017.
[36] F. A. Faria, P. Perre, R. A. Zucchi, L.R. Jorge, T. M. Lewinsohn, A. Rocha, and R. da S. Torres, Automatic identification of fruit flies (Diptera: Tephritidae)," *Journal of Visual Communication and Image Representation*, vol. 25, no. 7, pp. 1516–1527, Jul. 2014, doi: 10.1016/j.jvcir.2014.06.014.
[37] G. Tripathi and J. K. Save. "An image processing and neural network based approach for detection and classification of plant leaf diseases," *International Journal of Computer Engineering and Technology*, vol. 6, no. 4, pp. 14–20, 2015.

[38] D. M. Kabala, A. Hafiane, L. Bobelin, and R. Canals, "Image-based crop disease detection with federated learning," *Scientific Reports*, vol. 13, no. 1, Nov. 2023, doi: 10.1038/s41598-023-46218-5.

[39] I. N. Yulita, M. F. R. Rambe, A. Sholahuddin, and A. S. Prabuwono, "A convolutional neural network algorithm for pest detection using GoogleNet," *AgriEngineering*, vol. 5, no. 4, pp. 2366–2380, Dec. 2023, doi: 10.3390/agriengineering5040145.

[40] R. K. Dhanaraj, Md. A. Ali, A. K. Sharma, and A. Nayyar, "Deep multibranch fusion residual network and IoT-based pest detection system using sound analytics in large agricultural field," *Multimedia Tools and Applications*, vol. 83, no. 13, pp. 40215–40252, Oct. 2023, doi: 10.1007/s11042-023-16897-3.

[41] C. J. A. Bradshaw, B. Leroy, C. Bellard, D. Roiz, C. Albert, A. Fournier, M. Barbet-Massin, J-M Salles, F. Simard, and F. Courchamp, "Massive yet grossly underestimated global costs of invasive insects," *Nature Communications*, vol. 7, no. 1, Oct. 2016, doi: 10.1038/ncomms12986.

[42] J. Wang, C. Lin, L. Ji, and A. Liang, "A new automatic identification system of insect images at the order level," *Knowledge-Based Systems*, vol. 33, pp. 102–110, Mar. 2012, doi: 10.1016/j.knosys.2012.03.014.

[43] A. Hetzroni, V. Soroker, and Y. Cohen, "Toward practical acoustic red palm weevil detection," *Computers and Electronics in Agriculture*, vol. 124, pp. 100–106, Apr. 2016, doi: 10.1016/j.compag.2016.03.018.

[44] R. W. Mankin, R. D. Hodges, H. T. Nagle, C. Schal, R. M. Pereira, and P. G. Koehler, "Acoustic indicators for targeted detection of stored product and urban insect pests by inexpensive infrared, acoustic, and vibrational detection of movement," *Journal of Economic Entomology*, vol. 103, no. 5, pp. 1636–1646, Oct. 2010, doi: 10.1603/ec10126.

[45] B. Liu, Z. Hu, Y. Zhao, Y. Bai, and Y. Wang, "Recognition of pyralidae insects using intelligent monitoring autonomous robot vehicle in natural farm scene," *arXiv (Cornell University)*, Jan. 2019, doi: 10.48550/arxiv.1903.10827.

[46] F. Li and Y. Xiong, "Automatic identification of butterfly species based on HoMSC and GLCMoIB," *The Visual Computer*, vol. 34, no. 11, pp. 1525–1533, Aug. 2017, doi: 10.1007/s00371-017-1426-1.

[47] K. Espinoza, D. L. Valera, J. A. Torres, A. López, and F. D. Molina-Aiz, "Combination of image processing and artificial neural networks as a novel approach for the identification of Bemisia tabaci and Frankliniella occidentalis on sticky traps in greenhouse agriculture," *Computers and Electronics in Agriculture*, vol. 127, pp. 495–505, Jul. 2016, doi: 10.1016/j.compag.2016.07.008.

[48] M. Maharlooei, S. Sivarajan, S. G. Bajwa, J. P. Harmon, and J. Nowatzki, "Detection of soybean aphids in a greenhouse using an image processing technique," *Computers and Electronics in Agriculture*, vol. 132, pp. 63–70, Dec. 2016, doi: 10.1016/j.compag.2016.11.019.

[49] J. Chen, Y. Fan, T. Wang, C. Zhang, Z. Qiu, and Y. He, "Automatic segmentation and counting of aphid nymphs on leaves using convolutional neural networks," *Agronomy*, vol. 8, no. 8, p. 129, Jul. 2018, doi: 10.3390/agronomy8080129.

[50] Zhao, Y., Wang, Y., Wang, J., Hu, Z., Lin, F. and Xu, M., "GMM and DRLSE based detection and segmentation of pests: A case study," in *Proceedings of the 2019 4th International Conference on Multimedia Systems and Signal Processing*, pp. 62–66, 2019.

[51] A. Guarnieri, S. Maini, G. Molari, and V. Rondelli, "Automatic trap for moth detection in integrated pest management," *Bulletin of Insectology*, vol. 64, no. 2, pp. 247–251, 2011.
[52] M. Silveira and A. Monteiro, "Automatic recognition and measurement of butterfly eyespot patterns," *Biosystems*, vol. 95, no. 2, pp. 130–136, 2009.
[53] Y. Kaya, L. Kayci, and M. Uyar, "Automatic identification of butterfly species based on local binary patterns and artificial neural network," *Applied Soft Computing*, vol. 28, pp. 132–137, Dec. 2014, doi: 10.1016/j.asoc.2014.11.046.
[54] K. A. P. Siriwardena, L. C. P. Fernando, N. Nanayakkara, K. F. G. Perera, A. D. N. T. Kumara, and T. Nanayakkara, "Portable acoustic device for detection of coconut palms infested by Rynchophorus ferrugineus (Coleoptera: Curculionidae)," *Crop Protection*, vol. 29, no. 1, pp. 25–29, Oct. 2009, doi: 10.1016/j.cropro.2009.09.002.
[55] A. Gutiérrez, V. Ruiz, E. Moltó, G. Tapia, and M. Del Mar Téllez, "Development of a bioacoustic sensor for the early detection of Red Palm Weevil (Rhynchophorus ferrugineus Olivier)," *Crop Protection*, vol. 29, no. 7, pp. 671–676, Mar. 2010, doi: 10.1016/j.cropro.2010.02.001.
[56] B. Martin, S. M. Shaby, and G. Premi, "Studies on acoustic activity of red palm weevil the deadly pest on coconut crops," *Procedia Materials Science*, vol. 10, pp. 455–466, Jan. 2015, doi: 10.1016/j.mspro.2015.06.081.
[57] B. Martin and V. Juliet, "A novel approach to identify red Palm weevil on palms," *Advanced Materials Research*, vol. 634–638, pp. 3853–3857, Jan. 2013, doi: 10.4028/www.scientific.net/amr.634-638.3853.
[58] S. K. Bhoi et al., "An internet of things assisted unmanned aerial vehicle based artificial intelligence model for rice pest detection," *Microprocessors and Microsystems*, vol. 80, p. 103607, Dec. 2020, doi: 10.1016/j.micpro.2020.103607.
[59] K. Thenmozhi and U. S. Reddy, "Crop pest classification based on deep convolutional neural network and transfer learning," *Computers and Electronics in Agriculture*, vol. 164, p. 104906, Aug. 2019, doi: 10.1016/j.compag.2019.104906.

Chapter 11

Smart technologies for predicting the shelf-life of fruits and vegetables toward sustainable agriculture

Darshit Bansal, Abhay Singhal, Asmit Yadav, Shanu Sharma, Vinay Singh, and Abhishek Kumar Shukla

11.1 INTRODUCTION

As economic conditions improve, the demand for vegetables, fruits, meat, seafood, and other products continues to rise, indicating a highly favorable market outlook. However, food is also susceptible to corruption and spoiling, resulting in considerable waste and economic losses [1]. Consequently, the significance of food shelf-life for consumers and businesses is unequivocal. The Food and Agriculture Organisation of the United Nations reports that around one-third of all food produced for human consumption, or nearly 1.3 billion tons, is lost or wasted globally [1,2].

11.1.1 Food losses in fresh produce

The primary cause of food waste is the degradation and rotting of food during storage or preparation. Fruit and vegetable waste constitutes around 42% of total food waste due to their limited shelf-life and vulnerability to environmental conditions during storage and transportation. It is caused by inefficient harvesting, storage, and transport problems and overstocking within the retail market. In developing countries, most of these losses are post-harvest due to a lack of appropriate infrastructure and cold storage facilities [1]. On the other hand, waste in developed countries tends to be higher at the retail and consumer levels because of over-purchasing and aesthetic standards. People also contribute significantly by spoiling edible parts and storing food improperly. This causes enormous wastage of resources such as water, energy, and labor inputs in production. Some of the key reasons for losses in fresh produce are [2]:

- Not proper harvesting
- Lack of control over temperature
- Lack of proper handling of produced items
- Improper supply chain management

DOI: 10.1201/9781003494782-11

11.1.2 Key stages of food wastage

Starting from the production of food items in a farm till the processing of them at the consumer side, the wastage occurs at various stages. The contribution of different processes to food wastage is depicted in Figure 11.1 [1].

Farm Stage (11% Wastage):

- A large amount of produce is wasted at the harvesting stage because of wrong methods, poor timing for harvesting, and ineffective labor practices [1].
- Farm oversupply and failure to respond to market demand frequently translate into late-season edible produce waste.

Post-Harvest and Storage (8% Wastage):

- Lack of storage facilities, temperature control, and quality handling practices causes spoilage during transport and storage [2].
- Environmental conditions, including humidity and temperature changes, hasten spoilage of perishable foods such as fruits and vegetables.

Processing (1% Wastage):

- In food processing, inefficiencies in peeling, sorting, and packaging cause wastage that could be avoided.
- Cosmetic requirements cause rejection of fruits and vegetables that are still nutritionally good.

Retail and Stores (6% Wastage):

- Stores are unaware of the future shelf-life of fruits and vegetables and hence might store the produce in an improper way, resulting in spoilage before time, overstocking, or unnecessary wastage [1].

Figure 11.1 Breakdown of food waste from farm to consumer [1].

- Overstocking, poor inventory management, and storage in retail settings cause wasted produce.
- Produce rejected due to failure to meet cosmetic standards is another major contributor to wastage.
- Lack of certainty regarding the shelf-life of goods can lead to incorrect rotation or prioritization of stock, thus causing products to go bad prior to sale or consumption, compounding overall wastage.

Home and Restaurant (10% Wastage):

- Excess purchasing, incorrect storage, and misinterpretation of shelf-life at the consumer level cause maximum losses.
- Restaurants contribute by generating plate waste and ineffective portion sizes.

Rotting fruits and vegetables also produce methane, a strong greenhouse gas, which helps to fuel global warming. Providing solutions to such losses is instrumental to increasing food security, decreasing environmental damage, and giving supply chain sustainability [3].

11.1.3 Challenges and solutions

Food waste is a worldwide issue, and without early preventative actions, the level of food waste is projected to rise by as much as 70% by 2050. It also affects the environment in terms of land, climate change, biodiversity loss, freshwater, marine, and air pollution [1].

Shelf-life is a significant attribute of the agri-food supply chain that defines marketability, consumer acceptability, and food safety [4]. The shelf-life of fruits and vegetables depends on a number of intrinsic and extrinsic factors, such as maturity, firmness, water content, temperature, and bruising. Conventional shelf-life prediction methods depend significantly on visual examination or destructive laboratory analysis, which are subjective, time-consuming, and not suitable for mass production [5]. Thus, objective, automatic, and scalable systems need to accurately predict the shelf-life of produce. With the rise of smart technologies, a transformative shift is underway in how shelf-life is assessed, monitored, and predicted [6].

The food and agriculture sector is undergoing drastic change with the adoption of the most advanced technologies to enhance efficiency and sustainability. The sector contributes to nearly 10% of global GDP and is the backbone of most economies, especially in the developing world. Post-harvest losses, inefficiencies in the supply chain, and growing consumer demand for fresh, high-quality products put tremendous pressure on the stakeholders [7].

Advanced technology and innovations can help the circular economy. Known as Industry 4.0, the fourth industrial revolution has brought technology that improves manufacturing, assists food traceability, raises food safety and quality, lowers food waste creation, and allows total supply chain transparency from farm to consumer [7,8]. Many examples exist to support this, for instance, works on digital technologies and more particularly, smart sensors, Internet of Things (IoT), and nonthermal food processing, among others.

At every stage of the food supply chain, Industry 4.0 technologies can offer feasible and safe solutions as well as improve food sustainability and sustainable development objectives. Though there is no universal consensus on what technologies fall under Industry 4.0, most research views artificial intelligence (AI), IoT, smart sensors, big data, blockchain, and robotics as fundamental components of Industry 4.0 in agriculture and the food sector [9]. The area of machine learning (ML) and sophisticated image technologies also offers significant promise to offer a solution for the creation of precise shelf-life forecasting systems. Advanced ML models like convolutional neural networks (CNNs) can nowadays assess intricate patterns to glean relevant information from photos, which can then assist in forecasting the quality of food in terms of ripeness, disease detection, and general grading [10].

Furthermore, with the advancements in imaging technologies such as capturing infrared images, thermal images, high-resolution images, and so forth, researchers can utilize these images for the development of more precise food quality grading systems to further predict the shelf-life of the food [11]. For example, surface characteristics of food such as color and texture can be captured through high-resolution RGB images, whereas the internal characteristics such as temperature, ripeness indicators, and so forth can be captured through thermal imaging devices. Utilizing the proper power of ML models and imaging technologies, a precise and efficient shelf-life detection model can be successfully developed [12].

Motivated by the power of advanced technologies and the need of the era for the development of accurate shelf-life prediction models for the sustainable agriculture sector, this chapter is focused on providing in-depth details of the shelf-life prediction process. The chapter contains a discussion on the factors affecting the shelf-life and indicators for detecting the shelf-life. Various traditional ways of shelf line detection methods are also presented. Furthermore, the role of smart technologies for shelf-life detection is discussed, followed by a discussion on different types of ML, deep learning, and hybrid models. A comparative analysis of different existing works is also presented at the end of the chapter to provide a conclusive impact for the readers.

11.2 SHELF-LIFE PREDICTION: AN OVERVIEW

Food products, such as fruits and vegetables, deteriorate with time when stored. Consequently, there is a time frame within which these products are acceptable for use. This timeframe, referred to as shelf-life, is the time from manufacturing until the point at which the product is not acceptable for utilization [4]. Figure 11.2 depicts the spoilage of an apple. In essence, the shelf-life is the duration within which a food product sustains its important characteristics and remains acceptable for consumption, whether kept by retailers or consumers [12].

Industries are trying to achieve the longest possible shelf-life for products while keeping costs down and taking into account the realities of how these products are treated by distributors, retailers, and consumers. For example, supermarkets generally expect a product to have at least 75% of its shelf-life when it comes into their distribution sites. A short shelf-life typically results in customer dissatisfaction, complaints, and ultimately loss of reputation and sales for the brand [6,13].

11.2.1 Factors affecting the shelf-life

The shelf-life of a food item is determined by various factors, including the natural attributes of the product, environmental conditions for storage and distribution, as well as the packaging. Intrinsic factors, for example, pH values, water activity (aw), the occurrence of enzymes, microorganisms, and reactive substances, are impacted by the raw materials and process used. Extrinsic factors, on the other hand, encompass environmental conditions such as temperature, relative humidity, light, pressure, and mechanical stresses during handling and transportation. These conditions cause the product to deteriorate at an increased rate. Packaging is also significant, as it can affect extrinsic factors and thus the rate at which a product loses quality. It's very necessary to understand the factors in detail for better prediction of the shelf-life of a particular food [3].

Figure 11.2 Time-lapse of apple spoilage [13].

11.2.1.1 *Physiological and chemical factors*

After harvesting, the fruits and vegetables are still alive, after which the physiological and biochemical changes continue to take place. These internal changes are major reasons for their deterioration. Some of the key factors under this category of spoilage include [14]:

- **Respiration Effect:** Fruits and vegetables respire using oxygen and emit carbon dioxide. Higher respiration rates consume stored energy faster, and thus spoilage occurs quickly; for example, green vegetables and berries are highly respiring, which is the reason that they have short shelf lives.
- **Effect due to Ethylene Production:** Ripeness of fruits and vegetables occurs due to ethylene, which is a natural plant hormone. Climacteric fruits such as bananas, mangoes, and apples produce huge amounts of ethylene during ripening, and non-climacteric fruits like grapes and strawberries are more susceptible to external ethylene. Proper control of ethylene is required to further control the uneven ripeness throughout the fruit.
- **Loss of Water:** Heat causes loss of moisture and can even lead to dehydration, which causes wilting, shrivelling, and loss of firmness. Green vegetables and herbs that have a high surface area to volume ratio are most susceptible to this condition.
- **Microbial Effect:** Proliferation of the natural microflora on the surface of the produce can result in deterioration, discoloration, and even bring the food into unsound safety conditions.
- **Activities of Enzymes:** Softening, degeneration, and browning will be caused by the action of enzymes such as polyphenol oxidase and pectinase.

11.2.1.2 *Impact of environmental factors on the fruit shelf-life*

Temperature, humidity, wind, and atmospheric pressure are some of the environmental conditions that significantly affect the post-harvest quality and shelf-life of fruits and vegetables [15].

- **Temperature:** The temperature directly affects the respiration and other metabolic activities of the food. An increase in temperature increases these activities, which further causes the rapid decay of food. A proper maintenance of temperature is highly recommended, as for certain fruits, the low temperature also causes problems [14].
- **Relative humidity** is a significant element affecting water loss and deterioration. Elevated humidity levels can diminish evaporation while fostering the proliferation of moulds and bacteria, leading to deterioration. In contrast, low humidity hastens moisture depletion, resulting in

dehydration and desiccation. To achieve optimal storage, it is crucial to maintain an appropriate humidity level (about 85%–95% for most fruits and vegetables) to inhibit water loss and microbial proliferation [14,16].

- **Oxygen and CO_2 Levels:** The respiration process of food depends on the level of oxygen, which affects the deterioration of fresh produce, whereas being a respiratory inhibitor, the concentration of carbon dioxide slows down this process. The ability to monitor oxygen and CO2 levels in storage and transportation environments enables control of the ripening process and prevention of early spoilage, leading to increased freshness of perishable produce [15].
- **Light Exposure:** Exposure to light, especially ultraviolet (UV) light, can hasten ripening for certain fruits and vegetables. Also, extended exposure to light promotes photochemical decay of nutrients and flavor. Light exposure should be minimized using controlled environments or quality packaging to save the nutritional quality and flavor of produce, mitigating the chance of nutrient destruction and quality reduction [17].
- **Ethylene Gas:** Most fruits and vegetables give off ethylene, a natural plant hormone that accelerates ripening and aging. Overexposure to ethylene can lead to ripening and spoiling of vegetables too quickly. By controlling levels of ethylene gas and manipulating the storage environment appropriately, the ripening process can be slowed down, and shelf-life can be prolonged, thereby limiting food loss [14].
- **Wind and Airflow:** Proper airflow ensures the reduction of humidity and ethylene around fruits and vegetables to prevent spoilage. But high wind or inadequate airflow can physically injure the produce while enhancing dehydration. There is a need for proper ventilation to ensure that there are optimal conditions under which the fruits and vegetables stay fresh without getting damaged by too much moisture loss or physical injury [15].

By knowing and managing these environmental conditions, one can significantly limit the wastage and spoilage of fruits and vegetables. Close monitoring and optimization of conditions along the supply chain means produce lasts longer, resulting in less waste and a more sustainable food system.

11.2.2 Indicators for shelf-life detection

Predicting shelf-life involves identifying quantifiable factors that represent the physiological state of a given product. Freshness, ripeness, and decay are often evaluated using the following indicators:

- **Changes in Color:** Color changes such as greenish-yellow coloration of bananas and tomatoes turning red depict ripening (increasing ripening indices) and usually accompany an increase in the amounts of chlorophyll, anthocyanins, and carotenoids.

- **Texture and Firmness:** Softening in the texture of fruits and vegetables happens due to the breakdown of their cells. It can now be measured using non-destructive acoustic sensors or, even more advanced, a penetrometer.
- **Changes in Aromas:** Volatility of certain organic compounds indicates certain stages of aging or development in food products, which can be observed through gas sensors.
- **Soluble Solids Content:** Commonly referred to as SSC, which is often measured in Brix, it is a crucial indicator of the sugar level and sweetness and maturity of fruits like melons and grapes.
- **pH and Acidity:** Harvest maturity is often indicated by an increase in sugar content and a decrease in acidity with the progression of fruit ripening. Changes in pH can also monitor the optimal harvest and consumption windows.
- **Visual Signs of Spoilage:** Signs indicative of microbial and chemical in nature spoilage include mold growth, discoloration, and exudation (liquid leaking) as well as abnormal odors.

The detection and monitoring of these indicators through manual or smart technologies is highly necessary for the development of shelf-prediction models.

11.3 SHELF-LIFE PREDICTION: METHODS

Managing the quality, safety, and marketability of fresh produce requires accurate prediction of its shelf-life. It allows all actors in the supply chain—from growers to sellers—to make proper decisions regarding the harvesting, storage, transportation, and distribution. Over time, methods for predicting shelf-life have advanced from simple empirical estimation to complex predictive technological models. This essay describes the primary traditional, contemporary, and modern approaches used in estimating the shelf-life of fruits and vegetables [5,6].

11.3.1 Traditional methods for shelf-life prediction

Conventional methods of estimating the shelf-life of fruits and vegetables are based on age and are basic, yet still applicable for basic quality control. These are fundamentally subjective human observations and basic instruments to ascertain the freshness, maturity, and general status of the product. These techniques do not have the precision or the scalability of contemporary techniques, but offer a fast and cheap means to ascertain quality, particularly in small and medium-sized enterprises and settings where advanced technology is inaccessible.

Some of the most typical old ways are [18]:

- **Visual Examination:** Visual examination of observable signs of ripe, rot, or rot. For instance, bananas and apples ripen, and their color changes. Bananas become yellow from green, and apples become red or yellow based on variety. Exterior surfaces, bruising, or discoloration are usually utilized to determine the status of fruits, which foretell how long they will be fresh. Aside from color, size and shape changes may also be taken as indicators of aging or maturation. For instance, when cucumbers dehydrate, they will shrink or get muddy, and the tomatoes may become soft at the stem's edge. When testing fruits like tomatoes and peaches, an individual can squeeze the fruit to check for softness and strength, which are important signs of maturation or spoilage. Generally firmness is typically related to freshness, but excessive softness and creaks mean that the product has become overripe or begun to degrade [18,19,20].
- **Odor Examination:** Smell testing is commonly employed to detect spoilage. Odors like fermented and acidic odors are clear signals that the fruit has already reached its expiration date and can no longer be consumed. For instance, the strongly fermented odor of spoiled bananas and apples shortens the time because they already give off a sign of fermentation or development of microbial growth. It involves separating fruits and vegetables according to their appearance. These include obvious flaws like size, color, staining, cutting, cracking, and others. For instance, sorting of tomatoes will be selected to be more endurance when sorting tomatoes with skin and uniform colors but cracked and split ones are set aside since defects hasten decay. Sorting can also be utilized to approximate the maturity or ripeness of a commodity. Texture changes, wrinkling, or softening of the case or alteration of the color of the seeds may signal maturity. In some instances, split or cracked skins occur in mature tomatoes. This means that the fruit is too old to consume and is most likely to reach the end of durability [19].

Traditional methods tend to be quick, cost-effective, and effective but with troublesome constraints. They depend on human instinct, yet they can be inconstant and arbitrary, particularly in judging goods with alternative sources or conditions. Traditional methods do not identify latent faults or corruption, which is unseen and intangible at the surface level. In spite of these limitations, conventional technologies still play a crucial role in the evaluation of durability, particularly in small markets and where advanced tools and technologies are not available. Thus, these techniques are valuable supplements to a more contemporary and quantitative method for establishing the durability of fruits and vegetables.

11.3.2 Conventional methods

- **Destructive and Chemical Testing:** Chemical tests involve measurements of changes in chemical bonds in foods, like pH values, acids, and oxidized products. Through tracking these chemical changes over time, you are able to gauge the quality of your food and know its longevity. This kind of test is usually used for irregular shapes, like canned goods and packaged snacks. One of the most significant advantages of chemical testing is that it can deliver precise, objective, and quantitative information regarding the status of a food product. Nevertheless, it involves special equipment and expertise and might be more costly than other methods. Chemical testing might not detect any form of corruption. These methods utilize simple scientific principles to deliver more reproducible and more consistent measurements [21].

  Moisture Content: Water loss recording for measuring freshness.

  Firmness Tests: Penetrometers to test the texture of fruits like apples.

  pH Measurement: Measuring acidity to assess ripeness or spoilage.

  Sugar Content (Brix Test): Capturing sweetness as fruits ripen.

  Drawback: The Test sample cannot be reused.
- **Respiration Rate:** Fruit and vegetables (F&V) are living tissue and have high respiratory rates that play a significant role in determining durability. Respiration eliminates oxygen and expels carbon dioxide, resulting in the loss of nutrients. Climatic fruits like mangoes and bananas exhibit peaks of respiratory rates when they reach maturity, whereas non-permeable fruits have a steady respiratory rate. Altered or controlled air-mediated regulation of respiratory rates can slow down metabolic activities and enhance durability [18].
- **Microbiological Spoilage:** F&V is prone to microbial rot because it contains water and nutrients. Microorganisms invade wounds and wear, leading to deterioration without changing appearance. Sanitary handling, proper packaging, and storage conditions are needed in order to minimize microorganism growth and the loss of freshness [19].
- **Moisture Loss:** Water loss largely results in weight loss in F&V, wild, and structural breakdown. It impacts surface-to-volume ratio, skin characteristics, and environmental conditions like atmospheric humidity and temperature. Storage by maintaining proper packaging and high relative humidity (85%) could minimize this issue [21].
- **Ethylene:** Ethylene (C_2H_4), a fruit growth hormone, regulates aging and aging of fruits and plant varieties. Climatic fruits have been induced to ripen at a level of merely 0.1–10 ppm, while unwanted exposure triggers spoilage. Ethylene, even at later stages of non-permeable fruits, stimulates respiration and production. In our project, ethylene monitoring is employed to learn about durability and maximize

storage conditions that reduce spoilage and wastage. Early treatment of chromonic fruits prior to maturation will improve durability, whereas ethylene management in non-climate fruits will halt maturation. Through the tracking of ethylene mirrors in real time, you can minimize corruption by 25%. This will lead to fresh products and minimal waste. This technology plays a significant role in fighting food and supply chain sustainability [22].

Older technologies are usually quick, effective, and cost-effective, but not without strict limitations. They are based on human judgment, but they can be subjective and unreliable, particularly when assessing products from different sources or under different conditions. These approaches do not identify concealed defects or corruption that cannot be detected or sensed on the surface. Notwithstanding these disadvantages, conventional techniques are still essential in the estimation of durability, particularly in small markets and where advanced technology and tools are unavailable. These techniques are thus valuable additions to a more quantitative and new-age approach in determining the durability of fruits and vegetables. The methods are beneficial but tend to get absorbed by the inability to capture subjectivity, product variety changes, and environment variability in real time.

11.3.3 Modern approaches for shelf-life prediction

Present practices for determining shelf-life for fruit and vegetables have an intense concentration on non-destructive approaches to analyze the product without physically degrading or impairing its quality. Advanced image and sensing techniques, including RGB imaging, spectroscopy (hyperspectral, Raman, and NIR), thermal imaging, and X-ray imaging, are utilized in these techniques to scan internal and external features such as color, texture, moisture, and structure [11–13]. In contrast to the conventional destructive testing with sampling and cutting, the non-destructive methods keep the product intact for sale or additional processing. Machine learning algorithms, especially CNNs, are incorporated into these methods to process images or data collected, providing precise predictions for ripeness, spoilage, and shelf-life left [10,13]. These innovations not only increase accuracy but also render the process scalable and appropriate for real-time monitoring in industrial applications, minimizing food waste and enhancing supply chain efficiency.

- **Spectroscopy Techniques:** Spectroscopy methods are non-destructive processes that examine the interaction between electromagnetic radiation and fruits and vegetables in order to assess their biochemical and molecular characteristics. Spectroscopy measures important indicators like moisture, sugar levels, acidity, and internal defects that play a crucial role in shelf-life. Through the capture of chemical composition changes due to ripening or decay, spectroscopy gives detailed and accurate observations of produce quality. The following are three important sub-techniques that are commonly employed in shelf-life prediction [23]:

- **Near-Infrared (NIR) Spectroscopy:** NIR spectroscopy analyzes the absorption and reflectance of near-infrared light (700–2,500 nm) to quantify internal parameters such as water content, sugar concentration, and acidity. It is very efficient for monitoring the ripening of fruits and vegetables by identifying starch-to-sugar conversion and water loss. Although swift and non-invasive, its low penetration depth makes it most appropriate for small or thin-skinned produce [23,24].
- **Raman Spectroscopy:** Raman spectroscopy examines molecular vibration by measuring the scattering of monochromatic light. It is very sensitive to pigments like chlorophyll and carotenoids, which give information about ripeness and structural integrity. It, for example, identifies biochemical changes in cell walls while the produce is ripening or spoiling. It needs controlled environmental conditions to perform at its best and is costlier compared to others [24].

- **Hyperspectral Imaging (HSI):** HSI integrates spectroscopy and imaging for gathering spatial and spectral information over a range of wavelengths. HSI detects chemical compounds and early signs of spoilage such as fungal infection or bruising before they can be seen. HSI is superior to map moisture distribution and find changes in texture and composition that are subtle but have to undergo sophisticated data processing and are more expensive than less complex imaging techniques [11,12].
- **X-Ray Imaging:** X-ray imaging is an advanced, non-destructive technique that has become a key tool for forecasting the longevity of fruits and vegetables. Due to the application of electromagnetic waves with permeable energy, it offers precise internal structure data without damaging the sample. This technique is ideal for identifying concealed defects like bruises, cavities, cracks, and empty cavities, which are significant markers of corruption and quality degradation. Unlike surface-based methods, X-ray imaging analyzes the sound within fruits and vegetables and breaks through outer attributes like colors and textures to enable improved assessment. For instance, you can identify apples, avocados, and citrus bruises. In addition, X-ray imaging can effectively track structural changes in ripening [12]. When combined with machine learning algorithms, the applications of X-ray imaging are further enhanced. Folding (CNNs) and advanced models can analyze X-ray images to receive significant features like density changes and structural flaws [13]. These features enable automatic identification of the aging stage and precise identification of remaining durability. In addition, X-ray imaging for industrial production is centrally significant, enabling X-ray images to be utilized in automated sorting and sorting systems to look for quality products in the market [11].

Reading X-ray images also demands sophisticated software and technical expert knowledge, which can lead to operational complexity. However, through continuous innovations in imaging technology and machine learning, these issues can be overcome, rendering X-ray imaging less expensive and more adaptive to a wider array of applications. The capacity to reveal flaws, pursue structural alterations, and provide trustworthy forecasts in terms of longevity is a golden asset for contemporary agriculture and supply chain management. In combination with technological development that continually enhances ability, X-imaging is central to cutting food loss, optimizing storage and transport, and providing consumers worldwide with a regular, high-quality product supply [12].

- **Thermal Imaging:** Thermal imaging is a new, non-destructive technique that has transformed the measurement of fruit and vegetable durability. It involves the use of infrared technology to capture variations in thermal patterns and temperature, conveying crucial information regarding the physiological and structural integrity of the product. Technology is able to identify internal conditions like bruises, cavities, and cracks not visible to the human eye or conventional RGB imaging. Moreover, thermal imaging makes it possible to capture vital parameters that have a direct contribution to durability, tracking the maturation stage, and loss of loss and moisture [25].

  One of the key benefits of thermal imaging is its sensitivity to subtle temperature changes caused by metabolic activity like respiration and microbial growth. Such an activity is an indicator of ripeness and dryness or incipient rot. Thermal imaging, for instance, can track elevated respiration rates in climatic fruits like bananas and mangoes to categorize maturation stages and forecast remaining durability. It also aids you in identifying moisture losses in dense vegetables and root vegetables and identifying the quality decline stages. More sophisticated versions like the folding seller neural network (CNNs) analyze the thermal patterns to classify products once they have ripened or are spoiled, with the algorithm prediction estimating durability highly accurately. This pairing makes possible automated, scalable, real-time quality checking, and that is why thermal imaging is suitable for industrial purposes like sorting, classification, and cooling chamber management [26]. With many advantages, thermal imaging is not always challenging. For instance, interpreting sophisticated thermal data entails sophisticated image processing and machine learning. Nevertheless, ongoing technical advances bridge these barriers, rendering this technique more effective and accessible. As technology has improved, thermal imaging has become a valuable instrument for sustainable and optimized farming techniques to the advantage of farmers, traders, and consumers [25].

- **RGB Imaging:** RGB imaging is a non-conclusive, non-destructive way of estimating the fruit and vegetable shelf-life. It takes images of surface features like color, texture, and shape and offers excellent references to the onset of aging, quality, and corruption. RGB imaging is highly applicable for the mature classification task, since color change in fruits and vegetables like bananas, tomatoes, and strawberries is an effective indicator of freshness and decay. In addition, RGB imaging is able to detect decay via not just moisture loss and shrinkage of products like cucumbers and succulent vegetables, but also via visible defects like discoloration, mould, and wrinkles. One of the key benefits of RGB imaging is that it is relatively easy and affordable. Standard digital cameras and machine learning algorithms supply scalable applications for real-time monitoring and quality control within production environments. Machine learning models, such as folding fish networks (CNNs), improve the process by automatically deriving properties from RGB images and sorting fruits and vegetables following maturation or melting. This is achievable through precise prediction of durability, eradication of human error, precise prediction of waste reduction, and making supply chains efficient [27,28].

Nonetheless, RGB imaging has its drawbacks. Surfaces can be scrutinized, and internal injuries like bruises and cavities can be neglected. It also reacts to external conditions like illumination, which can alter the precision of color-based measurements. Nonetheless, such issues are addressed with ongoing developments in multimodal fusion, such as machine learning, techniques of data expansion, and hyperspectral imaging. RGB imaging is an effective and affordable way to foresee durability with optimal compromises between accuracy, scalability, and affordability. Its capability to help in decision making in real time turns out to be an important commodity for today's food supply chains and farming in eliminating waste, guaranteeing fresh goods to customers. The various types of durability forecast the advantages and disadvantages. A comparison of the various approaches is illustrated in Table 11.1.

Table 11.1 Comparative analysis of existing approaches

Criteria	*Traditional*	*Conventional*	*Modern*
Accuracy	Low, subjective	Moderate	High
Scalability	Limited, labor-intensive	Moderate	High
Cost	Low	Moderate	Varies (can be high initially)
Time	Immediate but imprecise	Requires sample preparation and testing	Fast with automation
Sample integrity	Fully intact	Partially destructive	Fully intact

11.4 SMART TECHNOLOGIES FOR SHELF-LIFE PREDICTION

Smart methods rapidly transform the monitoring and forecasting of fruit and vegetable shelf-life. Unlike conventional techniques based on static and ex-hoc assessment, smart technologies make possible real-time, non-invasive, and very accurate forecasting through the combination of records, data analysis, and automation. These technologies are particularly useful in complex, perishable supply chains where early identification of quality closures avoids losses, maximizes logistics, and maintains food safety.

Here you will find the most significant intelligent technology applied to foresee the life of fresh products.

- IoT technology played a crucial role in the real-time monitoring of environmental conditions influencing product durability. For instance, an IoT-based system incorporating temperature and moisture sensors was used to monitor memory environments. This enables early detection of rot and more precise prediction of durability [29].
- AI and ML models are capable of forecasting the shelf-life of different foods through the analysis of extensive data records, including environmental conditions and product attributes. These models are capable of recognizing intricate patterns and making better predictions than conventional approaches [30]. For instance, deep learning models were developed to forecast browning in fresh-cut salads. This is a novel quality of quality by examining the product's picture [31].

Overall, these smart technologies are capable of providing a wide range of durable predictions using real-time surveillance data and through advanced data analysis techniques.

11.5 MACHINE LEARNING APPROACHES FOR PREDICTING SHELF-LIFE OF FRUITS AND VEGETABLES

ML algorithms have revolutionized the assessment of fruits and vegetables. Such a process employs both structured and unstructured data to analyze endogenous and exogenous factors influencing product spoilage. From simple models to sophisticated deep learning models, ML techniques were employed to estimate durability with increased accuracy, scalability, and robustness. In this section, such processes are explained, including their evolution, which describe how the horizon of the future will expand.

11.5.1 Conventional machine learning for shelf-life prediction

Conventional machine learning models rely on predetermined attributes derived from generation functions like color, texture, weight, and biochemical characteristics.

- **Support Vector Machine (SVM):** SVMs were widely employed for the purpose of classification. Texture features, color histograms, and weight deviations are inputs to SVM classifiers to separate fresh products from largely decorated or spoiled classes [8].
- **Random Forests:** A robust ensemble algorithm used for multi-feature analysis of predictors of shelf-life. Features such as moisture content, respiration rate, and amounts of ethylene emissions are utilized to train the model [9].
- **K-Nearest Neighbours (K-NN):** This is a model that is beneficial for small data records and categorizes fruits on the basis of similar samples according to similarity. It is utilized for operations like recognition of levels of maturation and classification of periods of durability [7].
- **Naive Bayes Classifier:** Probabilistic model based on Bayes' theorem with independent properties. Good for fast and straightforward classification tasks, that is, the classification of fruits based on pre-defined displacement measurements.
- **K-Means Clustering:** An unsupervised algorithm for splitting data into groups based on similarity and segregating fruits by the ripeness level or identifying spoilage patterns.

11.5.2 Deep learning models for shelf-life prediction

Deep learning (DL) has really promoted shelf-life prediction by making feature extraction automatic and dealing with massive amounts of data. The models work very well when used together with image technologies such as RGB and thermal imaging [28].

- **CNNs:** CNNs have been widely utilized for extracting spatial features from fruit images. Features, including color gradients, surface texture, and bruising, are recorded utilizing spectrogram-like representations of RGB and thermal imagery [27].
- **Recurrent Neural Networks and Long Short-Term Memory Networks (LSTMs):** These models learn temporal patterns in sequence data, for example, variation in respiration rates or temperature as a function of time [9].
- **Transformer Models:** Recent breakthroughs such as Vision Transformers (Vites) are being investigated for predicting shelf-life by processing sequential image data [10].

11.5.3 Hybrid and emerging models

Shelf-life detection hybrid models merge various methods, for example, machine learning algorithms, imaging techniques, and environmental sensors, to improve prediction accuracy. These models bring together the advantages of DL, for example, CNNs, and analytical methods like regression or Bayesian methodologies to handle different types of data. Hybrid models apply RGB, thermal, and hyperspectral images to external and internal quality testing, with environmental conditions such as temperature, humidity, and ethylene concentrations to estimate shelf-life. By fusing information from multi-modal sources, they achieve a more holistic and accurate prediction. Such models are superior in addressing uncertainty and providing flexibility for different conditions across supply chains. They are better than individual models by enhancing robustness and scalability, with a great reduction in food wastage. Hybrid methods are best suited for real-time tracking, enabling stakeholders to have actionable insights to maximize storage and distribution networks for perishable commodities [10].

- **Transfer Learning:** Transfer learning is a machine learning process where a model is developed for a task to carry out related but distinct tasks. Rather than training the model from the start, transfer learning leverages the knowledge within an already developed network. This is typically trained on large and diverse data sets. It is achieved by fine-tuning the model by rewriting its parameters with smaller domain-specific data sets. For instance, trained models can be adjusted in ImageNet, a big image database, to label medical images and interpret fruit durability from visual abilities. Lightweight prefabricated CNN models like SqueezeNet, MobileNETV2, and Shuffle-Net were utilized in literature for fruit classification based on the available fresh levels [10]. Thermal imaging was also employed to analyze crucial characteristics like interior textures, bruising, and other quality factors, creating a non-destructive and useful way of evaluating durability [25]. This approach greatly minimizes computational burden and time from the training model's point of view right from the beginning, but ensures high precision. The study shows the significance of structured models in agricultural methods, which means that models requiring minimal retraining can be adapted to particular problems like fruit classification and durability estimation. Thermal imaging performed better when capturing internal quality characteristics compared to RGB imaging. The study shows that the Shuffle-Net works best for application in practice from among the tested models, being efficient and very accurate. Through the combination of thermal imaging and transfer learning, studies offer sound and scalable technologies for opportunities of reducing postharvest losses. The framework allows for the possibility of enhancing the efficiency and sustainability of agricultural resources.

- **Bayesian Neural Networks (BNNs):** BNNs provide a probabilistic solution for predicting durability with uncertainty. In contrast to point estimates for regular neuronal networks, the distribution of BNN parameters (weights) is predicted, allowing for the prediction of the size associated with prediction uncertainty. This is especially useful when predicting fruit durability. For instance, one can estimate that the mangoes belong to the "RUL-5" category, which has a 70% probability of a chance, and the "RUL-4" category, which has a 30% probability of a chance [30]. A probabilistic version like this makes way for sensible decisions. For instance, if the image has inconsistent lighting or if the fruit is jammed in different environmental conditions, BNN can actually incorporate this uncertainty into its output. This capability can be utilized to make prediction systems more robust, resilient to real variations, and prevent potential misclassification and subsequent food waste.
- **Reinforcement Learning (RL):** RL can be applied to build a process for establishing the remaining useful life of a fruit. RL differs from monitoring learning with agents when labeled data records are utilized, trying out environments, and learning to get rewards or penalties for feedback. As for the matter of retention life, RL agents can test fruit-related knowledge, including intrinsic properties like texture, color, and heat behavior, as well as extrinsic environmental parameters like temperature and air humidity within the storage environment. The RL approach gives conditions, actions, and rewards their specifications. The condition may be a present observable attribute of the fruit. Actions can forecast intervention suggestions like particular types of durability ("remaining useful lifespan of 5 days") or optimizing storage conditions. Rewards are made up to encourage accurate predictions and penalize mistakes. For instance, agents can accurately predict that there will be mangoes and reward them if they penalize them for underestimating freshness [26]. With time, the decisions can be made optimal not only for estimating durability but also for suggesting steps to minimize food wastage. This is a dynamic and quick-response strategy, especially when information is limited or when the system must be generalized under a wide range of conditions, such as varying types of fruit and storage conditions. By incorporating RL, it complements other transfer learning methods to give a strong framework for surmounting the complexity of predicting durability [32].

11.6 COMPARISON OF EXISTING APPROACHES

A lot of work has been done in the literature utilizing different ML approaches for automatic food grading and shelf-life prediction. ML models proved to be highly efficient in tackling some issues of predicting fruit and vegetable longevity. One such usage is that ML models like CNN are capable of recognizing levels of maturity through color changes and texture modification with over 90% precision [7]. For instance, strawberries or tomatoes are grouped based on the evolution of the color, a good indicator of freshness, to further classify them through a CNN. Another major application is forecasting the shelf-life of fruits and vegetables, that is, bananas and avocados. LSTM networks have been utilized, which evolve over time by exploring temporal variation, that is, respiratory rates and ethylene release. These techniques are used not only for the prediction of the duration of durability, but also for enhancing general quality control in farm practices.

The combination of DL models and IoT sensors allows for real-time monitoring of fruit quality. Real-time data, such as temperature, humidity in the air, and light pollution, can be combined with visual data in order to enhance durability estimation. In a study, the capability of thermal imaging combined with CNN models was tested for real-time prediction of fruit quality [13]. This technique is highly applicable in cold rooms where observation of real time is necessary to achieve the freshness of stored plants. The details of some of the existing works are summarized in Table 11.2.

11.7 CONCLUSION AND FUTURE DIRECTIONS

Shelf-life prediction of fruits and vegetables remains crucial for minimizing post-harvest waste, optimizing supply chain efficiency, and addressing worldwide food security concerns. This chapter has assessed a variety of machine learning techniques, ranging from classic models, including support vector machines and random forests, to more sophisticated deep learning architectures such as CNNs and long short-term memory networks. These approaches, combined with nondestructive sensing technologies such as RGB and thermal imaging, have exhibited strong potential to automate analyses of ripeness and deterioration. Novel methods such as transfer learning and generative adversarial networks have also addressed data limitations and improved the scalability of solutions. While innovative methods have yielded higher accuracy and efficiency, traditional models remain applicable in resource-constrained settings. This integration paves the way for future advances toward multimodal data integration and developing stable, scalable systems for shelf-life estimation. Through these technologies, the agricultural sector can progress toward achieving sustainable and efficient produce management, drastically reducing food waste and optimizing resource usage.

Table 11.2 Review of existing related work on machine learning-based shelf-life prediction

Ref	*Approach*	*Task*	*Dataset/features*	*Methodology*	*Performance*	*Analysis*
[33]	SVM	Ripeness classification	RGB images, texture-based features	SVM classifier with RBF kernel for separating ripe vs. unripe classes.	85% accuracy on mango datasets	Effective for binary classification; requires careful feature engineering.
[34]	Random Forest	Shelf-life prediction	Temperature, humidity, and moisture content	Random Forest with feature importance analysis to predict shelf-life duration.	Achieved 88% acc. for strawberries under controlled conditions	Robust to noisy data; interpretable results but computationally intensive.
[35]	CNN	Ripeness and spoilage detection	RGB images for color and texture, thermal images for internal defects	Deep CNN architecture with convolutional and pooling layers for feature extraction.	>95% accuracy on mango and banana datasets	Excellent at capturing spatial patterns but requires large labeled datasets and GPU resources.
[36]	Transfer Learning	Shelf-life prediction	Pre-trained MobileNetV2 fine-tuned on fruit datasets	Fine-tuning of pre-trained models	92% accuracy with minimal labeled data	Reduces the need for extensive datasets; risk of overfitting if not tuned correctly.
[37]	LSTM	Temporal modeling of ripening	Time-series data (temperature, ethylene levels)	Sequential analysis of changes over time to model ripening and spoilage patterns.	Achieved >90% accuracy for avocado datasets	Effective for time-series data; high computational cost and memory usage.

REFERENCES

1. R. Porat, A. Lichter, L. A. Terry, R. Harker, and J. Buzby, "Postharvest losses of fruit and vegetables during retail and in consumers' homes: Quantifications, causes, and means of prevention," *Postharvest Biology and Technology*, vol. 139, pp. 135–149, Feb. 2018, doi: 10.1016/j.postharvbio.2017.11.019.
2. E. M. Karoney, T. Molelekoa, M. Bill, N. Siyoum, and L. Korsten, "Global research network analysis of fresh produce postharvest technology: Innovative trends for loss reduction," *Postharvest Biology and Technology*, vol. 208, p. 112642, Nov. 2023, doi: 10.1016/j.postharvbio.2023.112642.
3. S. Duret et al., "Identification of the significant factors in food quality using global sensitivity analysis and the accept-and-reject algorithm. Part I: Methodology," *Journal of Food Engineering*, vol. 148, pp. 53–57, Oct. 2014, doi: 10.1016/j.jfoodeng.2014.09.037.
4. R. Torres-Sánchez, M. T. Martínez-Zafra, N. Castillejo, A. Guillamón-Frutos, and F. Artés-Hernández, "Real-time monitoring system for shelf life estimation of fruit and vegetables," *Sensors*, vol. 20, no. 7, p. 1860, Mar. 2020, doi: 10.3390/s20071860.
5. B. Fu and T. P. Labuza, "Shelf-life prediction: Theory and application," *Food Control*, vol. 4, no. 3, pp. 125–133, Jan. 1993, doi: 10.1016/0956-7135(93)90298-3.
6. A. Abougharib, M. Awad and M. Ndiaye, "Remaining shelf-life estimation of fresh fruits and vegetables during transportation," *IEEE Access*, vol. 11, pp. 8845–8859, 2023, doi: 10.1109/ACCESS.2023.3239584.
7. M. Palumbo et al., "Emerging postharvest technologies to enhance the shelf-life of fruit and vegetables: An overview," *Foods*, vol. 11, no. 23, p. 3925, Dec. 2022, doi: 10.3390/foods11233925.
8. D. Li, L. Bai, R. Wang, and S. Ying, "Research progress of machine learning in extending and regulating the shelf life of fruits and vegetables," *Foods*, vol. 13, no. 19, p. 3025, Sep. 2024, doi: 10.3390/foods13193025.
9. T. Srikanth, A. Poornima, D. Balamani, and C. Sadha, "Estimating the remaining shelf-life using CNN SLEM of fresh fruit and vegetables while transportation," *International Journal of Engineering in Computer Science*, vol. 6, no. 2, pp. 62–66, Jul. 2024, doi: 10.33545/26633582.2024.v6.i2a.127.
10. S. Tammina, "Transfer learning using VGG-16 with deep convolutional neural network for classifying images," *International Journal of Scientific and Research Publications*, vol. 9, no. 10, p. p9420, Oct. 2019, doi: 10.29322/ijsrp.9.10.2019.p9420.
11. Y. Lu, W. Saeys, M. Kim, Y. Peng, and R. Lu, "Hyperspectral imaging technology for quality and safety evaluation of horticultural products: A review and celebration of the past 20-year progress," *Postharvest Biology and Technology*, vol. 170, p. 111318, Aug. 2020, doi: 10.1016/j.postharvbio.2020.111318.
12. B. Zhang et al., "Detection of early rottenness on apples by using hyperspectral imaging combined with spectral analysis and image processing," *Food Analytical Methods*, vol. 8, no. 8, pp. 2075–2086, Jan. 2015, doi: 10.1007/s12161-015-0097-7.
13. Y. Li, X. Feng, Y. Liu, and X. Han, "Apple quality identification and classification by image processing based on convolutional neural networks," *Scientific Reports*, vol. 11, no. 1, Aug. 2021, doi: 10.1038/s41598-021-96103-2.

14. U. E. Umeohia and A. A. Olapade, "Physiological processes affecting postharvest quality of fresh fruits and vegetables," *Asian Food Science Journal*, vol. 23, no. 4, pp. 1–14, Apr. 2024, doi: 10.9734/afsj/2024/v23i4706.
15. L. Wang and M. Teplitski, "Microbiological food safety considerations in shelf-life extension of fresh fruits and vegetables," *Current Opinion in Biotechnology*, vol. 80, p. 102895, Jan. 2023, doi: 10.1016/j.copbio.2023.102895.
16. M. Kafkaletou, A. Velliou, M. V. Christopoulos, G. Ouzounidou, and E. Tsantili, "Impact of cold storage temperature and shelf life on ripening physiology, quality attributes, and nutritional value in apricots—Implication of cultivar," *Plants*, vol. 12, no. 15, p. 2875, Aug. 2023, doi: 10.3390/plants12152875.
17. I. K. Arah, H. Amaglo, E. K. Kumah, and H. Ofori, "Preharvest and postharvest factors affecting the quality and shelf life of harvested tomatoes: A mini review," *International Journal of Agronomy*, vol. 2015, pp. 1–6, Jan. 2015, doi: 10.1155/2015/478041.
18. A. Valero, E. Carrasco, and R. Ma, "Principles and methodologies for the determination of shelf-life in foods," in *Trends in Vital Food and Control Engineering*, InTech *eBooks*, 2012, pp. 3–42. doi: 10.5772/35353.
19. R. P. Singh, "Scientific principles of shelf life evaluation," in *Shelf Life Evaluation of Foods*, Springer eBooks, 1994, pp. 3–26. doi: 10.1007/978-1-4615-2095-5_1.
20. D. S. Prabha and J. S. Kumar, "Assessment of banana fruit maturity by image processing technique," *Journal of Food Science and Technology*, vol. 52, no. 3, pp. 1316–1327, Oct. 2013, doi: 10.1007/s13197-013-1188-3.
21. Y. Phimolsiripol and P. Suppakul, "Techniques in shelf life evaluation of food products," in *Reference Module in Food Science*, Elsevier eBooks, 2016, pp. 1–8. doi: 10.1016/b978-0-08-100596-5.03293-5.
22. T. M. E. Tellez, R. Quevedo-León, and O. Diaz-Carrasco, "Un indicator values for food shelf life prediction: A review," *Scientia Agropecuaria*, vol. 15, no. 3, pp. 429–448, Jul. 2024, doi: 10.17268/sci.agropecu.2024.032.
23. D. Tejerina, R. Contador, and A. Ortiz, "Near infrared spectroscopy (NIRS) as tool for classification into official commercial categories and shelf-life storage times of pre-sliced modified atmosphere packaged Iberian dry-cured loin," *Food Chemistry*, vol. 356, p. 129733, Mar. 2021, doi: 10.1016/j.foodchem.2021.129733.
24. D. Eisenstecken, A. Panarese, P. Robatscher, C. W. Huck, A. Zanella, and M. Oberhuber, "A near infrared spectroscopy (NIRS) and chemometric approach to improve apple fruit quality management: A case study on the cultivars 'Cripps Pink' and 'Braeburn,'" *Molecules*, vol. 20, no. 8, pp. 13603–13619, Jul. 2015, doi: 10.3390/molecules200813603.
25. P. Pathmanaban, B. K. Gnanavel, S. S. Anandan, and S. Sathiyamurthy, "Advancing post-harvest fruit handling through AI-based thermal imaging: Applications, challenges, and future trends," *Discover Food*, vol. 3, no. 1, Dec. 2023, doi: 10.1007/s44187-023-00068-2.
26. B. Yang, Y.-C. Hung, G. D. Kumar, K. Casulli, and K. M. Solval, "From sunburn detection to optimal cooling: A review of recent applications of thermal imaging to improve preharvest and postharvest handling of fruit and vegetables," *Scientia Horticulturae*, vol. 337, p. 113527, Jul. 2024, doi: 10.1016/j.scienta.2024.113527.

27. A. S. M. F. Al Haque, M. A. Hakim and R. Hafiz, "CNN based automatic computer vision system for strain detection and quality identification of banana," in *2021 International Conference on Automation, Control and Mechatronics for Industry 4.0 (ACMI)*, Rajshahi, Bangladesh, 2021, pp. 1–6, doi: 10.1109/ACMI53878.2021.9528269.
28. A. Rastogi, R. Arora and S. Sharma, "Leaf disease detection and grading using computer vision technology & fuzzy logic," in *2015 2nd International Conference on Signal Processing and Integrated Networks (SPIN)*, Noida, India, 2015, pp. 500–505, doi: 10.1109/SPIN.2015.7095350.
29. S. Hazra, A. Mondal, P. Dey, S. Prabhakar, A. K. Jha and N. Rakshit, "Future prospects of agriculture using iot and machine learning," in *2024 IEEE International Conference on Smart Power Control and Renewable Energy (ICSPCRE)*, Rourkela, India, 2024, pp. 1–6, doi: 10.1109/ICSPCRE62303.2024.10674786.
30. J. Dutta, P. Deshpande, and B. Rai, "AI-based soft-sensor for shelf life prediction of 'Kesar' mango," *SN Applied Sciences*, vol. 3, no. 6, May 2021, doi: 10.1007/s42452-021-04657-7.
31. N. Aherwadi, U. Mittal, J. Singla, N. Z. Jhanjhi, A. Yassine, and M. S. Hossain, "Prediction of fruit maturity, quality, and its life using deep learning algorithms," *Electronics*, vol. 11, no. 24, p. 4100, Dec. 2022, doi: 10.3390/electronics11244100.
32. C. Shi et al., "Artificial neural network-based shelf life prediction approach in the food storage process: A review," *Critical Reviews in Food Science and Nutrition*, vol. 64, no. 32, pp. 12009–12024, Sep. 2023, doi: 10.1080/10408398.2023.2245899.
33. D. Worasawate, P. Sakunasinha, and S. Chiangga, "Automatic classification of the ripeness stage of mango fruit using a machine learning approach," *AgriEngineering*, vol. 4, no. 1, pp. 32–47, Jan. 2022, doi: 10.3390/agriengineering4010003.
34. M. N. Ribeiro et al., "Quality control of fresh strawberries by a random forest model," *Journal of the Science of Food and Agriculture*, vol. 101, no. 11, pp. 4514–4522, Jan. 2021, doi: 10.1002/jsfa.11092.
35. J. Han, T. Li, Y. He, and Q. Gao, "Using machine learning approaches for food quality detection," *Mathematical Problems in Engineering*, vol. 2022, pp. 1–9, Dec. 2022, doi: 10.1155/2022/6852022.
36. V. Bhole and A. Kumar, "A transfer learning-based approach to predict the shelf life of fruit," *Inteligencia Artificial*, vol. 24, no. 67, pp. 102–120, Apr. 2021, doi: 10.4114/intartif.vol24iss67pp102-120.
37. Y. J. Davur, W. Kämper, K. Khoshelham, S. J. Trueman, and S. H. Bai, "Estimating the ripeness of hass avocado fruit using deep learning with hyperspectral imaging," *Horticulturae*, vol. 9, no. 5, p. 599, May 2023, doi: 10.3390/horticulturae9050599.

Chapter 12

Predictive analytics in agriculture

Machine learning for crop yield estimation

Siddhi Dingorkar, Pranali Kosamkar, Alok Bajoria, Yukta Shah, Ankit Dahiya, and Mahi Bohara

12.1 INTRODUCTION

Traditional farming has been transformed by the application of machine learning (ML) in agriculture, which has addressed issues of food security and sustainability [1]. Farmers, legislators, and other stakeholders are better equipped to make decisions when crop yield forecasts and recommendations are accurate. For instance, ML has enhanced agricultural health monitoring in India and effectively optimized irrigation systems in California [2]. These developments lessen problems like climate fluctuation and resource mismanagement, which are major obstacles in contemporary agriculture [3].

Conventional techniques are susceptible to shifting weather patterns and excessive fertilizer use because they mostly rely on gut feeling and past patterns. ML contributes to global food security by ensuring more robust, effective, and sustainable farming practices through the use of historical data and sophisticated algorithms [4, 5].

12.2 DATA AND PREPROCESSING

12.2.1 Data collections

Mendeley and Kaggle datasets are used in the study:

Suggestion for Crops Features include state, district, crop year, season, crop type, area cultivated, and production (tons) in 240,692 rows of the dataset.

Forecast for Crop Yield: 2,201 rows make up the dataset, which includes information on temperature, humidity, pH, rainfall, soil nutrients (N, P, and K), and suggested crops.

Limitations:

Although the datasets offer a wealth of information, their lack of geographical diversity and real-time updates may restrict the models' usefulness in other areas [6].

DOI: 10.1201/9781003494782-12

12.2.2 Preprocessing steps

Steps in preprocessing include:

- Data cleaning:
  Handled missing values using mean/median imputation for numerical attributes and mode for categorical attributes [7].
  The IQR approach was used to identify outliers in characteristics like as production and area cultivated.
- Standardization and Normalization:
  Standardized characteristics such as rainfall, N, P, and K are used to guarantee consistent feature contribution. Applied z-score normalization [8].
- Feature Selection:
  Retained highly associated characteristics with the target variable using PCA, boosting model performance.
- Data Splitting:
  Divide the data into 80:20 training and testing sets. The data were used for testing and validation, and the rest was used to train the models.
  High-quality data input for precise predictions is ensured by this methodical pretreatment.
  By performing data cleaning, normalizing, and selecting features, we ensure that the data fed into our models is of good quality, thereby improving the accuracy and reliability of our crop yield forecasting and recommendations [9,10].

12.3 METHODOLOGY

The system architecture (Figure 12.1) illustrates the workflow, integrating data preprocessing, algorithm implementation, and model evaluation.

Algorithms Used
Crop Yield Prediction: Linear Regression, Decision Tree Regression, and Random Forest Regression.
Crop Recommendation: Random Forest, KNN, SVM, and Decision Tree.

For instance, KNN classifies crops by comparing feature similarities with neighboring data points, achieving optimal accuracy through parameter tuning (e.g., selecting an appropriate value for "k").

Justification for the Algorithm Choice

The selected algorithms balance accuracy, interpretability, and computational efficiency, making them suitable for diverse agricultural tasks.

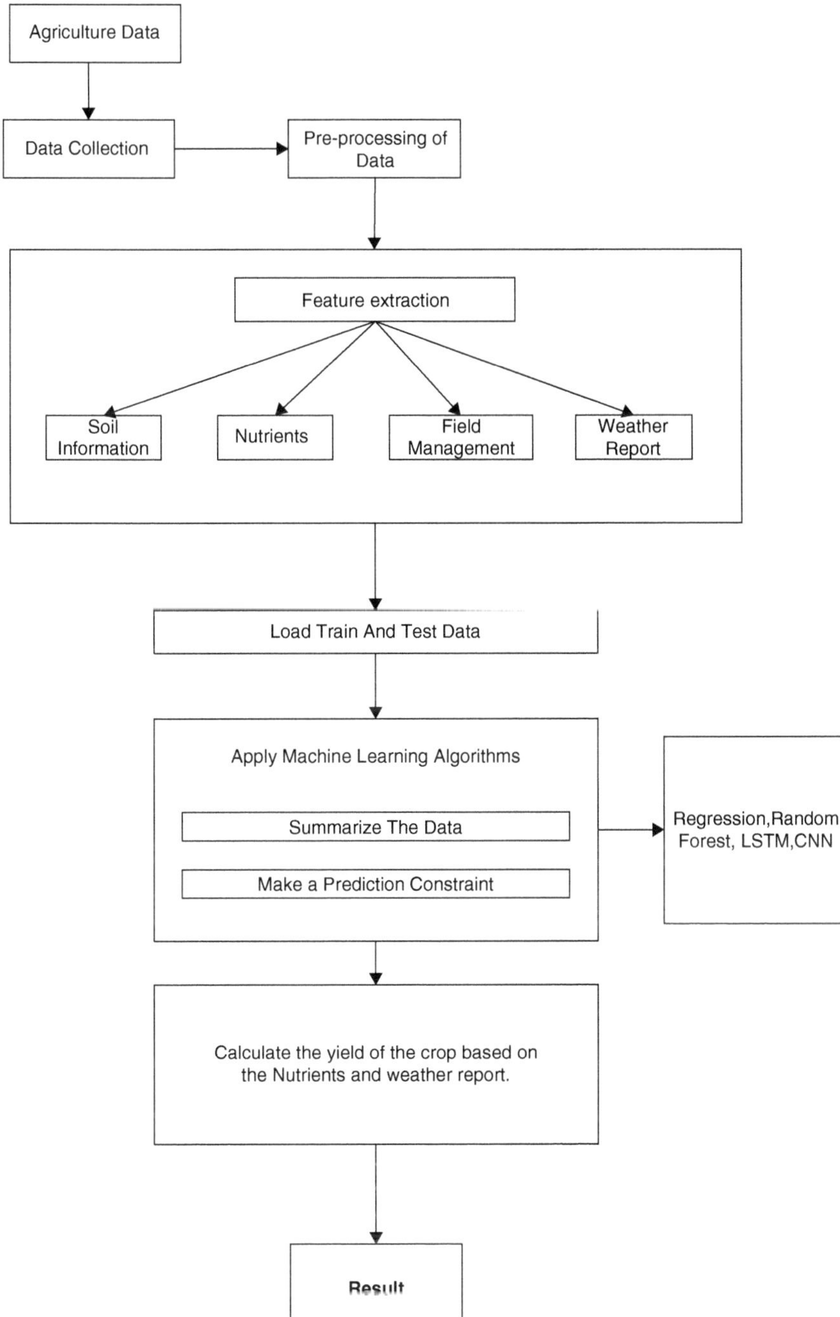

Figure 12.1 System architecture for the proposed system.

12.3.1 Algorithms used

Regression algorithms, namely, Linear Regression, Decision Tree Regression, and Random Forest Regression, are utilized to forecast crop yield. Variables including State Name, District Name, Crop Year, Season, Crop, Area, and Production are included in the yield prediction dataset. These variables give the models the vital information they require to be trained and classification techniques, namely, Random Forest, K-Nearest Neighbors (KNN), Support Vector Machine (SVM), and Decision Tree, for crop recommendation. Features like soil pH, soil nutrient levels (N, P, and K), and weather conditions (temperature, humidity, and rainfall) are included in the dataset for crop recommendation. These characteristics are vital because they have a big impact on crop productivity and suitability [11, 12].

- KNN: The KNN model uses the resemblance between a crop's features and those of nearby data points to classify crops. For example, KNN endorses the crop best appropriate for those conditions if a data point's neighboring data points have similar quantities of nutrients (N, P, and K) [13, 14] .

  Getting the Best Accuracy with KNN: By adjusting the value of k, which controls the number of closest neighbors taken into account in the prediction, we were able to achieve the best accuracy with the KNN classifier [15]. A modest choice of k produced the most dependable results, balancing the model's sensitivity to noise while guaranteeing accurate classification, according to our thorough experimentation and cross-validation. The KNN model for crop classification functions is given in Figure 12.2.
- SVM: The SVM model divides several crop classes in the feature space by identifying the best hyperplane. It offers accurate crop recommendations by assisting in the classification of crops into discrete categories according to their environmental and nutritional qualities [16, 17].
- Linear Regression: This algorithm simulates the link between the input features (State Name, District Name, Crop Year, Season, Crop, and Area) and the crop yield (Production). Because it makes the assumption that characteristics and yield follow a linear relationship, it can be used to forecast crop yields using historical data [18, 19].
- Decision Tree Regression: This approach divides the dataset into smaller subgroups based on feature values, much like the decision trees used for classification. The result is a tree-like structure. A yield prediction is represented by each leaf node, which captures feature interactions and non-linear correlations [20, 21].
- Random Forest Regression: The yield is predicted using an ensemble of decision trees. The final prediction is the average of all the trees' yield projections, each of which is based on a subset of features [22, 23].

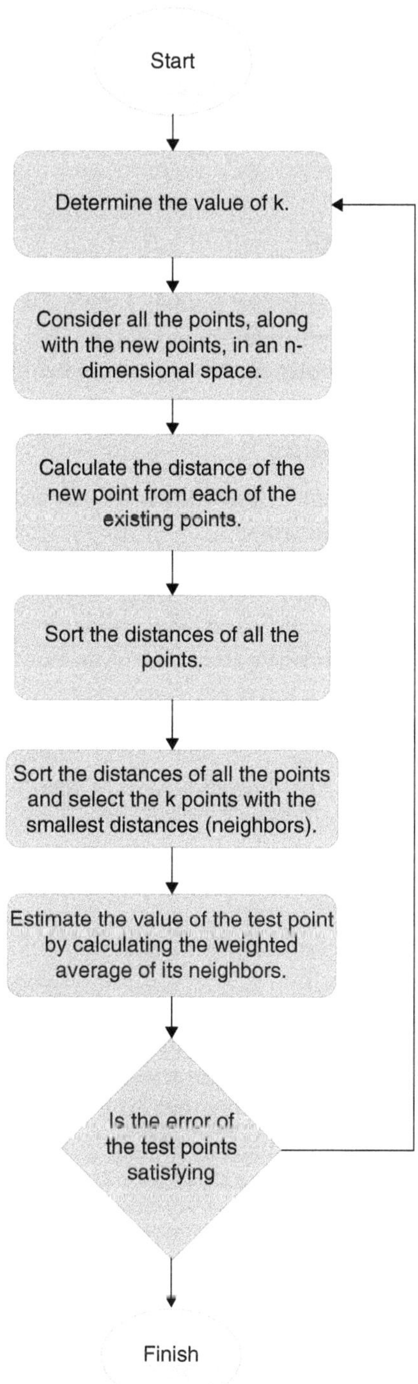

Figure 12.2 Working of the KNN classification model.

12.3.2 Performance metrics

We used the metrics like accuracy, precision, recall, and F1-score for crop recommendation classification systems. Measures like Mean Absolute Error (MAE), Mean Squared Error (MSE), and Root Mean Squared Error (RMSE) are used for regression methods used in yield prediction [24, 25].

12.4 MODEL BUILDING

Reliable and broadly applicable ML models need to be built using a systematic process that includes multiple crucial stages. We examine each of the major steps in the model implementation process below.

12.4.1 Algorithm building

We have selected a wide range of algorithms for our study according to how well they perform particular tasks:

- **Crop Recommendation:** Based on their efficacious classification and recommendation of crops based on a range of agricultural and environmental criteria, Decision Trees, Random Forests, SVM, and KNN were chosen.
- **Crop Yield Prediction:** Based on their ability to represent intricate relationships in the data, Random Forest Regression, Decision Tree Regression, and Linear Regression were selected to forecast crop yields [24].

12.4.1.1 Data preparation

The data preparation essential actions consist of:

- Feature engineering is the process of turning unprocessed data into useful features that improve model performance.
- **Data Splitting:** The dataset is split into training and testing sets an 80-20 split is employed, meaning that 80% of the data are used to train the models and the remaining 20% are set aside for accuracy testing.

12.4.1.2 Model training

By minimizing the discrepancy between expected and actual values, models are trained using the training dataset.

- Using the training dataset, each model is trained, and its parameters are optimized to lower error.

- Cross-validation is used to evaluate the performance of the model on unknown data. The training data are divided into k subgroups for k-fold cross-validation. The model is trained k times, using one validation set for each subset. By doing this, overfitting is reduced and the model is fine-tuned.

12.4.1.3 Model evaluation

The testing dataset is used to assess the models once they have been trained.

- Metrics, including accuracy, precision, recall, and F1-score, are utilized for classification jobs. MAE, MSE, and RMSE are calculated for regression tasks.
- To increase the models' predictive capacity, additional model tweaking and refining may be done in light of the evaluation's findings.

12.4.2 Training and testing

12.4.2.1 Data splitting

Data splitting separates the data into two major groups:

- **Training Set:** This set, which makes up 80% of the dataset, is what the ML models are trained on.
- **Testing Set:** The models' capacity to generalize to previously unseen data are assessed using the final 20% of the dataset.

12.4.2.2 Cross-validation

K-fold cross-validation is used to verify the models' dependability and prevent overfitting:

- The model is trained and verified k times, using a different subset as the validation set for each of the k equal subsets that were created from the training data. A reliable estimation of the model's performance is produced by this procedure.

Metrics of Performance:

Depending on the type of task, various metrics are used:

- The main metrics utilized for classification models (crop recommendation) are accuracy, precision, recall, and F1-score.
- Metrics like MAE, MSE, and RMSE are computed to assess performance for regression models (crop yield prediction).

12.4.3 Hyperparameter tuning

We employed the Grid Search, Random Search, and Bayesian Optimization for hyperparameter tuning.

12.4.3.1 Hyperparameters for every model

For the models in this study, the following are the important hyperparameters:

- **KNN:** K is the number of neighbors.
- **SVM:** Kernel type (e.g., linear, polynomial, Radial Basis Function (RBF)), kernel coefficient (gamma), and regularization parameter (C).
- **Decision Tree:** Maximum depth, minimum sample split, and minimum sample leaf.
- **Random Forest:** Minimum sample split, minimum sample leaf, maximum depth, and number of trees (n_estimators).
- **Linear Regression:** Regularization parameters, particularly when using regularized versions like Ridge or Lasso.
- **Decision Tree Regression:** Maximum depth, minimum sample split, and minimum sample leaf.
- **Random Forest Regression:** Number of trees (n_estimators), maximum depth, minimum sample split, and minimum sample leaf.

12.4.3.2 Process

- **Initial Model Training:** Set the default hyperparameters for each model and start training.
- **Hyperparameter Search:** To explore the hyperparameter space, use Bayesian optimization, random search, or grid search.
- **Model Evaluation:** Utilize cross-validation to assess the model's performance with several sets of hyperparameters.
- **Pick the Ideal Model:** Select the model configuration that produces the best results during cross-validation.
- **Last Instruction:** Utilizing the complete training dataset, optimize the hyperparameters of the selected model.
- **Testing:** Lastly, evaluate the model's performance to confirm its efficacy using the testing dataset.

These procedures help us make sure that our ML models are optimized and fine-tuned for precise yield estimates and crop recommendations, which in turn help to promote more sustainable and knowledgeable agriculture practices.

12.5 RESULTS AND DISCUSSIONS

The correlation matrix (Figure 12.3) identifies relationships among variables such as temperature, humidity, and rainfall. For instance:

- **Temperature and Humidity:** A weak positive correlation (0.19) indicates that there is a slight tendency for humidity to rise with temperature.
- Temperature and rainfall have a very weak correlation (0.03), suggesting that there is no meaningful relationship.

  By directing feature selection and model improvement, these insights improve prediction performance.

12.5.1 Crop year against area, yield, and production

Analyzing the relationship between crop year and factors like as yield, area, and production offers important information on long-term trends. A negative correlation points to possible declines that need more research, while a positive correlation points to an upward tendency.

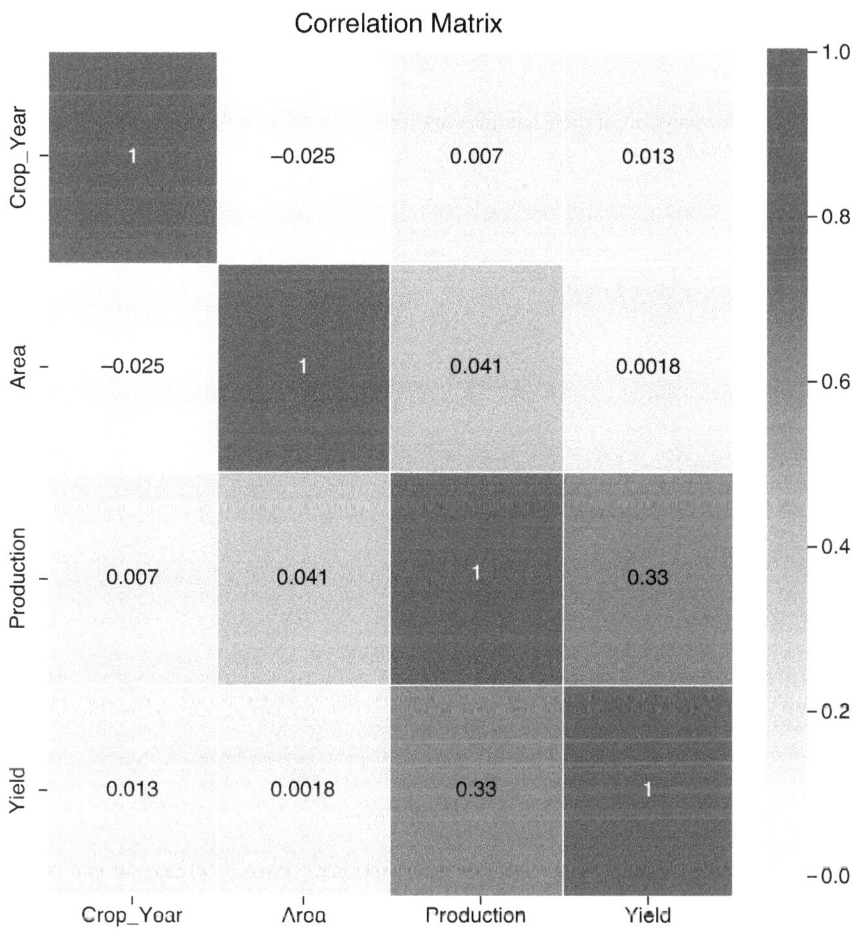

Figure 12.3 Correlation matrix of crop yield prediction.

12.5.2 Random Forest Regression

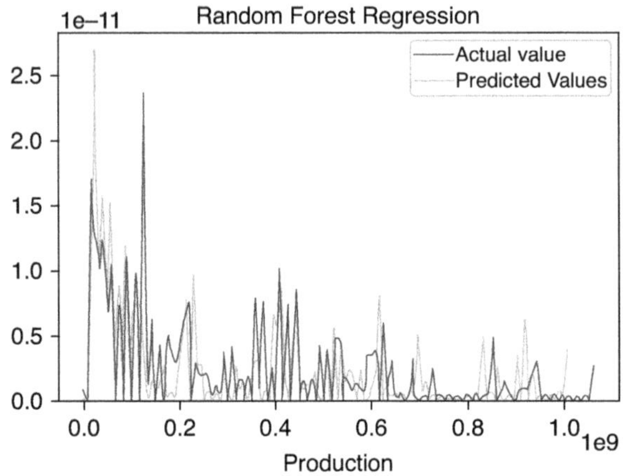

Figure 12.4 Comparison of the Random Forest Regression model's predicted and actual values.

12.5.3 Decision tree

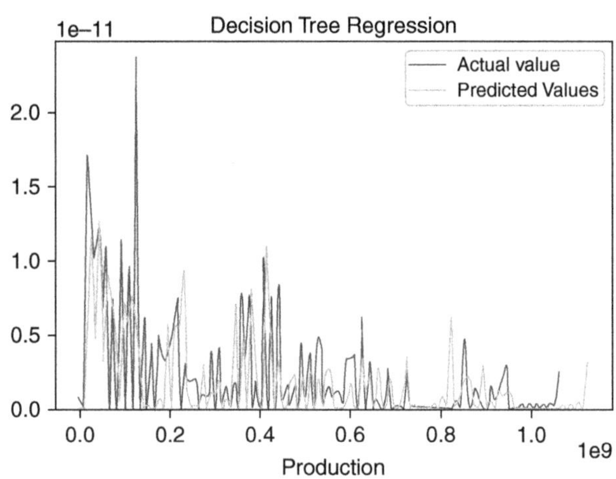

Figure 12.5 Comparison of the decision tree regression model's predicted and actual values.

12.5.4 Linear regression

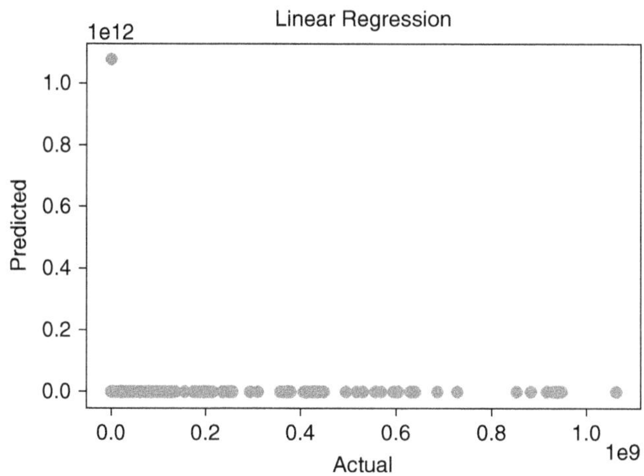

Figure 12.6 Scatter plot of actual values for the linear regression model compared to predicted values.

12.5.5 Model performance

Random Forest Regression: Achieved the highest accuracy of 96.67%.

Figure 12.4 shows a strong alignment between predicted and actual values, indicating high reliability.

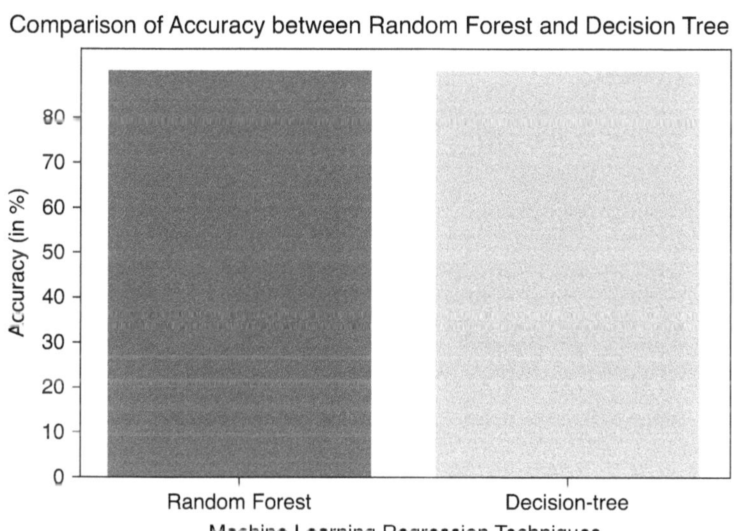

Figure 12.7 The comparison of accuracy between the decision tree and Random Forest Regression models.

Decision Tree Regression: Balanced accuracy and computational efficiency with an accuracy of 92.43% (Figure 12.5).

Linear Regression: Demonstrated suboptimal performance due to its linear assumptions, with an accuracy of −66,175.59 (Figure 12.6).

Insights from Underperformance: Linear Regression's struggles with nonlinear relationships emphasize the importance of advanced algorithms like Random Forest and Decision Tree for complex datasets.

With respect to Figure 12.7, it can be observed that the Random Forest Regression and Decision Tree Regression models demonstrate noteworthy accuracy scores of 0.9625 and 0.9564, respectively. The performance of each in yield prediction is shown in the comparison graph above.

The Crop Recommendation System dataset was analyzed using a number of classifiers, including Random Forest, Decision Tree, SVM, and KNN. Their performance was evaluated not only on accuracy but also on standard deviation.

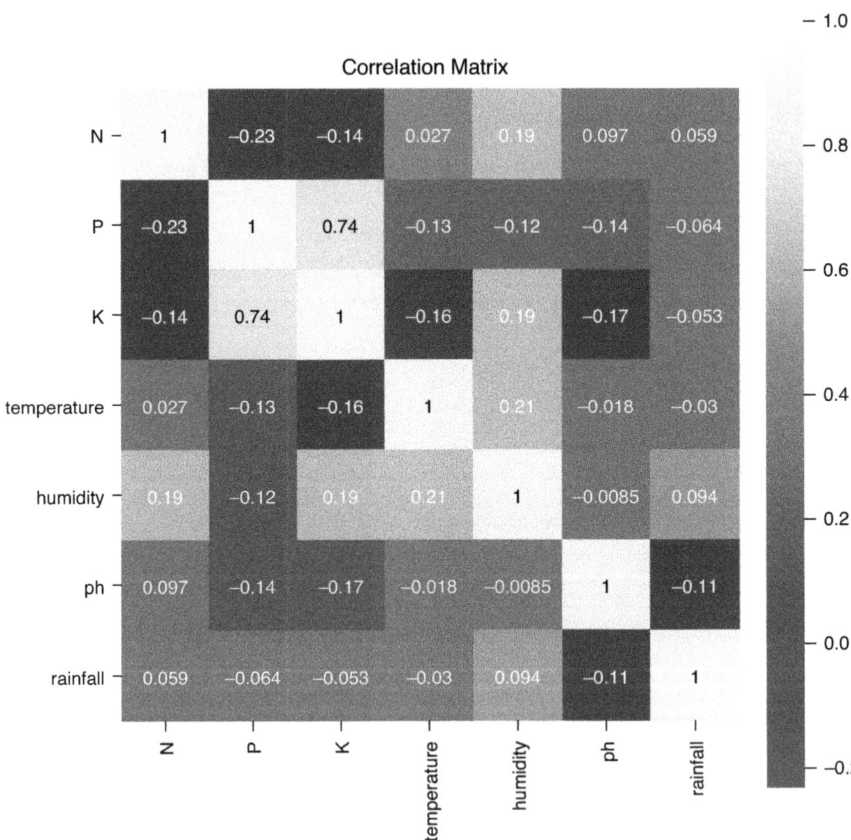

Figure 12.8 Correlation matrix for the crop recommendation system.

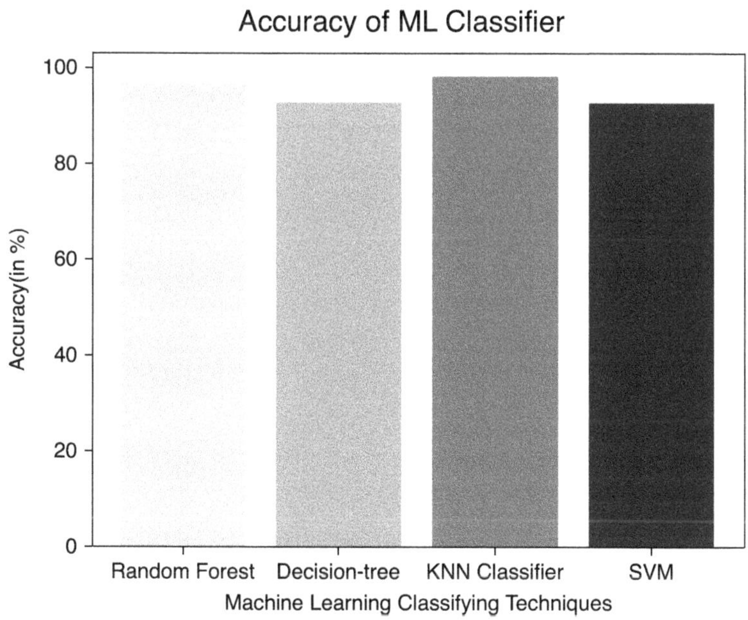

Figure 12.9 Bar graph showing the accuracy of various classification techniques vs accuracy.

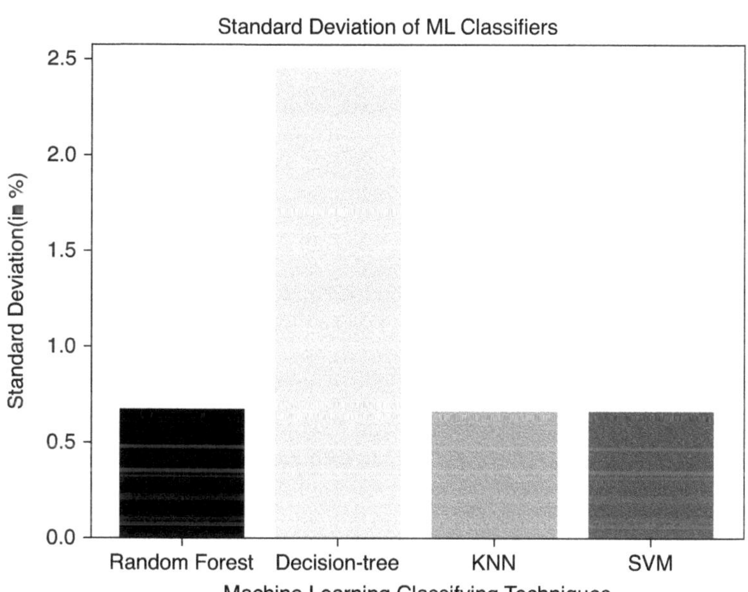

Figure 12.10 Bar graph showing the accuracy of various classification techniques vs standard deviation.

Table 12.1 Machine learning classification techniques

Sr. no.	*Algorithm*	*Accuracy*	*Standard deviation*
1	Random Forest	96.67	0.691
2	Decision tree	92.43	2.520
3	KNN classifier	98.18	0.668
4	SVM	90.54	0.657

Table 12.2 Machine learning regression techniques

Sr. no.	*Algorithm*	*Accuracy*	*F2 score*
1	Linear regression	−66,175.59	−66,175.59
2	Random Forest	0.9643	0.9643
3	Decision tree	0.9585	0.9585

As depicted in Figure 12.8, a weakly positive correlation (around 0.19) exists between humidity and temperature. This suggests that humidity may have a slight, weak propensity to rise in response to temperature.

The link between rainfall and temperature is rather small (around 0.03), implying that there is basically no relationship at all between these two variables. Additionally, there is a marginally negative correlation (around −0.12) between rainfall and humidity, suggesting a tiny tendency for rainfall to slack off as humidity increases.

Figure 12.9 visually compares the accuracies of these models, confirming KNN's superior performance for classification tasks.

Figure 12.10 compares the standard deviations, revealing that the Decision Tree exhibits the highest variation, which may affect its consistency in predictions.

Tables 12.1 and 12.2 show the comparison of the ML model with accuracy, standard deviation, and F2 score.

12.6 CONCLUSION

In this chapter, we study how various ML algorithms, KNN, SVM, and RF, are used in predicting crop yields and recommending suitable crops. The research results show that RF achieved an impressive accuracy of 96.67%, SVM reached 90.54%, and KNN attained 98.18%, underscoring their potential for real-world application. Such high accuracy levels indicate that these models can effectively support farmers in planning their crop strategies and improving yield outcomes. Future research could focus on refining these models further to enhance prediction accuracy and suggestion precision. This could involve exploring new features, integrating additional data sources, and leveraging advanced ML techniques. By continually improving these models, we can better support sustainable and productive agricultural practices.

REFERENCES

[1] M. Champaneri et al., "Crop yield prediction using machine learning," *Technology*, vol. 9, no. 38, 2016.

[2] D. J. Reddy and M. R. Kumar, "Crop yield prediction using machine learning algorithm," *2021 5th International Conference on Intelligent Computing and Control Systems (ICICCS)*, Madurai, India, 2021, pp. 1466–1470, doi: 10.1109/ICICCS51141.2021.9432236.

[3] D. A. Bondre and S. Mahagaonkar, "Prediction of crop yield and fertilizer recommendation using machine learning algorithms," *International Journal of Engineering Applied Sciences and Technology*, vol. 4, no. 5, pp. 371–376, 2019.

[4] R. Medar, V. S. Rajpurohit, and S. Shweta, "Crop Yield Prediction using Machine Learning Techniques," *2019 IEEE 5th International Conference for Convergence in Technology (I2CT)*, Bombay, India, 2019, pp. 1–5, doi: 10.1109/I2CT45611.2019.9033611.

[5] M. Suganya, R. Dayana, and R. Revathi, "Crop yield prediction using supervised learning techniques," *International Journal of Computer Engineering and Technology*, vol. 11, no. 2, pp. 9–20, 2020.

[6] M. Rashid, B. S. Bari, Y. Yusup, M. A. Kamaruddin, and N. Khan, "A comprehensive review of crop yield prediction using machine learning approaches with special emphasis on palm oil yield prediction," *IEEE Access*, vol. 9, pp. 63406–63439, 2021, doi: 10.1109/ACCESS.2021.3075159.

[7] S. Khaki and L. Wang, "Crop yield prediction using deep neural networks," *Frontiers in Plant Science*, vol. 10, p. 621, 2019, doi: 10.3389/fpls.2019.00621.

[8] D. Ramesh and B. V. Vardhan, "Analysis of crop yield prediction using data mining techniques," *International Journal of Research in Engineering and Technology*, vol. 4, no. 1, pp. 47–473, 2015.

[9] S. Khaki, L. Wang, and S. V. Archontoulis, "A CNN-RNN framework for crop yield prediction," *Frontiers in Plant Science*, vol. 10, p. 1750, 2020, doi: 10.3389/fpls.2019.01750.

[10] P. Muruganantham, S. Wibowo, S. Grandhi, N. H. Samrat, and N. Islam, "A systematic literature review on crop yield prediction with deep learning and remote sensing," *Remote Sensing*, vol. 14, no. 1990, 2022, doi: 10.3390/rs14091990.

[11] P. Gopi and M. Karthikeyan, "Multimodal machine learning based crop recommendation and yield prediction model," *Intelligent Automation & Soft Computing*, vol. 36, pp. 313–326, 2023, doi: 10.32604/iasc.2023.029756.

[12] R. Pawar et al., "The Role of Innovative Data Mining Approaches for Analyzing and Estimating the Crop Yield in Agriculture among Emerging Nations," *2022 International Conference on Advances in Computing, Communication and Applied Informatics (ACCAI)*, Chennai, India, 2022, pp. 2339–2342, doi: 10.1109/ICACITE53722.2022.9823729.

[13] A. Crane-Droesch, "Machine learning for crop yield prediction," *Environmental Research Letters*, vol. 13, no. 114003, 2018, doi: 10.1088/1748-9326/aae159
[14] A. Nyéki and M. Neményi, "Crop Yield Prediction in Precision Agriculture," *Agronomy*, vol. 12, no. 10, p. 2460, 2022, doi: 10.3390/agronomy12102460.
[15] Y. M. Fernandez-Ordoñez and J. Soria-Ruiz, "Maize Crop Yield Estimation with Remote Sensing and Empirical Models," *2017 IEEE International Geoscience and Remote Sensing Symposium (IGARSS)*, Fort Worth, TX, USA, 2017, pp. 3035–3038, doi: 10.1109/IGARSS.2017.8127638.
[16] R. A. Arun et al., "A Machine Learning-based Agricultural Yield Forecasting System for Predicting Crop/Plant Yield before Planting the Crops/Plants," *2023 12th International Conference on Advanced Computing (ICoAC)*, Chennai, India, 2023, pp. 1–6, doi: 10.1109/ICoAC59537.2023.10249364.
[17] A. Sharma et al., "Early Prediction of Crop Yield in India Using Machine Learning," *2022 IEEE Region 10 Symposium (TENSYMP)*, Mumbai, India, 2022, pp. 1–6, doi: 10.1109/TENSYMP54529.2022.9864490.
[18] A. Lakshmanarao et al., "Crop Yield Prediction Using Regression Models in Machine Learning," *2023 2nd International Conference on Applied Artificial Intelligence and Computing (ICAAIC)*, Salem, India, 2023, pp. 423–426, doi: 10.1109/ICAAIC56838.2023.10141462.
[19] P. S. Bharathi et al., "An Experimental Analysis of Crop Yield Prediction Using Modified Deep Learning Strategy," *2022 International Conference on Advances in Computing, Communication and Applied Informatics (ACCAI)*, Chennai, India, 2022, pp. 1–6, doi: 10.1109/ACCAI53970.2022.9752492.
[20] S. Thirumal and R. Latha, "Automated Hyperparameter Tuned Stacked Autoencoder Based Rice Crop Yield Prediction Model," *2023 7th International Conference on Trends in Electronics and Informatics (ICOEI)*, Tirunelveli, India, 2023, pp. 14–18, doi: 10.1109/ICOEI56765.2023.10126037.
[21] A. Ghosh et al., "A Comprehensive Crop Recommendation System Integrating Machine Learning and Deep Learning Models," *2024 1st International Conference on Cognitive, Green and Ubiquitous Computing (IC-CGU)*, Bhubaneswar, India, 2024, pp. 1–6, doi: 10.1109/IC-CGU58078.2024.10530724.
[22] Ed-daoudi, Rachid, et al. "A predictive approach to improving agricultural productivity in morocco through crop recommendations." *International Journal of Advanced Computer Science and Applications*, vol. 14, no. 3, 2023. doi: 10.14569/IJACSA.2023.0140322.
[23] A. Banerjee, S. Paul, and S. Bhattacharya, "Advanced Agricultural Systems: Identification, Crop Yields, and Recommendations Using Image-Processing Techniques and Machine-Learning Algorithms," in *Machine Learning, Blockchain, and Cyber Security in Smart Environments*, Chapman and Hall/CRC, 2022, pp. 81–97.
[24] T. Kalpana et al., "An Investigation and Forecast of Suitable Crop Yield Using Machine Learning Techniques Dependent on the Characteristics of Soil," in *2024 International Conference on Advances in Computing, Communication and Applied Informatics (ACCAI)*, IEEE, 2024.
[25] Y. J. Praveen et al., "Machine Learning-Based Hybrid System for Rainfall Prediction and Crop Recommendation," in *2024 International Conference on Advances in Modern Age Technologies for Health and Engineering Science (AMATHE)*, IEEE, 2024.

Chapter 13

AI-driven wild animal detection and tracking in agriculture fields using deep learning

Showkat A. Dar, T. Saravanan, R. Indhumathi, Mithra C, and Aafaq A. Rather

13.1 INTRODUCTION

Agriculturalists worldwide face various complex problems that result from the simultaneous existence of farming and wildlife. One of these is when wild animals invade fields; as a result, they damage crops and become a source of monetary loss [1]. As previously stated, this relationship between people and animals can grow into a full-fledged conflict that negatively impacts farming production and wildlife conservation efforts [2]. It is critical to implement successful animal identification and monitoring approaches on farms to combat this issue.

There have been many years of challenges in detecting and watching wild animals in agro landscapes [3]. Traditional methods, such as human observation or physical obstacles, are often time-consuming, expensive, and fragile, as they need to efficiently deter animals from restricted areas due to the highly dynamic behavior of the animals and the vast areas of agricultural lands [4]. Additionally, identifying and treating invaders must be swift, and it must be done with non-invasive and ecofriendly practices.

Furthermore, aside from the issues discussed above, there is also a rise in the awareness of the sustainability requirement. The latter is crucial for eliminating conflicts between animals and people in farming domains [5]. Commonly used approaches to deter animals, including fencing and using chemical repellents, harm wildlife and the environment [6]. As a result, innovative, environmentally friendly, innovative, innovative technological solutions that can efficiently solve the abovementioned issues are required.

The use of deep machines helps to alleviate the problem of identifying and monitoring wild animals in farms. It involves the application of deep machine learning (ML), a form of ML that employs neural networks comprising multiple layers to decode complex patterns and associations from data. Scientists can use such models to model sophisticated prediction methods that can accurately and efficiently detect and monitor wild animals within farms [7].

One of the applications of deep learning in identifying wild animals is the analysis of remote sensing data; it may include satellite imagery or drone

DOI: 10.1201/9781003494782-13

recordings. Indeed, this bird's-eye view of agricultural fields can easily detect animal intrusions [8]. Subsequent analysis of recorded data by deep learning may identify animals as separate objects from other elements, thus detecting wildlife presence automatically.

Another solution is to use sensor networks with cameras and various environmental sensors across agricultural areas. In particular, these sensors allow for monitoring animal activity, such as movement or behavior. Thus, the data are obtained much faster, and deep learning algorithms can rapidly analyze it to immediately identify and track any potential threats posed by wild animals to agriculture and act accordingly by farmers or wildlife managers. In conclusion, combining deep learning with remote sensing and sensor networks addresses the problem of wildlife detection and monitoring in agricultural areas. It ensures a balance between agriculture and wildlife conservation. This chapter was provided with a structure of introduction in this section, followed by literature in Section 13.2, the proposed system in Section 13.3, experimental results in Section 13.4, and conclusion.

13.2 LITERATURE

Bhagabati et al. [9] highlighted creating a system for real-time detection and classification of wild animals in video feeds and notifications to prevent avoidable interactions between humans and animals. The system detects wild animals in real time by implementing deep learning models and YoloV5 with the SENet attention layer. To shorten the time used for the manual labeling process creation and adequate trial and training data, it was conducted by blending open-source and bespoke datasets from various animal species. A cloud-based artificial intelligence system was implemented in the cameras to take photos from various KNP areas and validate the model's efficacy.

Rančić et al. [10] presented many state-of-the-art network architectures that have been trained on a well-annotated picture collection to predict the existence of objects in the dataset. Three permutations of the You Only Look Once and a Single Shot Multibox Detector architecture identify deer in heavily wooded environments. The efficacy of these models' effector information, including mean average precision, recall, and F1 score, was reviewed. These models are also assessed based on their proficiency in real time.

Mao et al. [11] presented a series of research concerning Animal Activity Recognition (AAR) using wearable sensors and deep learning algorithms. The text outlines sensor kinds often used and animal species and behaviors commonly examined. Moreover, the text gives an exhaustive examination of deep learning technologies utilized in wearable sensor-enhanced AARs. The division of technologies is made according to the classification of deep learning algorithms. The following paragraphs provide a wide range of publicly available datasets produced over the past five years using both wearable and augmented reality sensors. This knowledge would be beneficial for potential researchers. Further, the article discusses the possible difficulties of deploying a deep learning model in the subject area. Additionally, the article presents the alternatives and suggests the follow-up research. The analysis conducted herein provides the reader with feasible outcomes in enhancing the current AAR systems' efficiency through wearable and wearable sensors. The automatic system, which gathers measurable results combined with the veterinarian professional's subject-view, shows high potential in enhancing the species' health and well-being.

Fernando et al. [12] suggested a comprehensive approach to deter animals by considering many characteristics such as color, coat pattern, morphology, and diurnal and nocturnal vocalizations. Moreover, it can detect the number of animals approaching crops and monitor animal behavior to prevent inaccurate alerts. For these objectives, the research uses several methodologies, such as image processing and deep learning, to examine aural, visual, and pictorial data obtained from particular animal populations.

Natarajan et al. [13] presented a novel model that integrates the VGG-19 architecture with a Bi-LSTM network. The model is designed for animal identification systems and alert generation based on detected activities. The systems ensure that Short Message Service is sent to the nearest forest office for rapid responses. The proposed model has shown more progress in general performance.

Sajithra Varun et al. [14] presented a new method to identify animals, which is shown by employing new advancements in Deep Learning algorithms. The proposed solution is implemented by utilizing deep learning technology to spot animals early on and take security measures to eliminate harm due to animals. This work integrates image processing and artificial intelligence techniques to spot animals, recognize their species, and identify them automatically. CNNs are utilized for image recognition purposes. It also includes an alert module and animal-repellent circuitry. The proposed methodology experimented with datasets that have been meticulously created for animal validation, including Amur Tiger Re-identification in the Wild, Animal Detection Image Datasets, and Google Open Images V6+. Deep CNNs are utilized to extract data from intricate optical pictures. Typical camera trap databases are used to create hybrid CNN models. After the classification process, the retrieved CNN features are input into highly efficient deep learning algorithms. As a result of this process, it was observed that the performance was of better quality than several other publications.

Sayagavi et al. [15] developed a novel, efficient, and reliable computer vision method for the automatic identification of wild animals. The technique utilizes the YOLO object detection paradigm to detect the existence of untamed creatures in photos. More precisely, the approach is designed to identify six specific categories: humans and five separate species of animals (elephant, zebra, giraffe, lion, and cheetah). Once animals are detected, their movements are tracked using CSRT to determine their intent. Based on this information, alerts are sent to notify the appropriate authorities. In addition, the article describes the creation of a prototype for this suggested approach, using Raspberry Pi devices that are outfitted with cameras.

Thomas et al. [16] suggested a solution by combining the Internet of Things (IoT) with ML approaches. Advancements in image processing and IoT sensor networks have led to significant sensor advances. The problem of animal-human conflicts in agricultural areas and forest zones is substantial, endangering human lives and resulting in significant financial losses. A wireless sensor-based animal incursion detection system may be created by analyzing video clips from a given dataset. The current study presents the utilization of the SlowFast architecture for Invariant Feature Extraction. Video annotation is performed, and spatial data are extracted while utilizing the IFE model. A novel type of monitoring may be conducted if the animals are classified by their characteristics based on the visual analysis. It is feasible to lower animal-vehicle collisions, improve animal monitoring, and prevent theft via detection and classification methods. The usage of deep learning methods is vital for effective implementation.

13.3 PROPOSED METHOD

Proposing the integration of the Unit with EfficientNet-B7 for wild boar semantic segmentation is a significant innovation in computer vision. The integration of the robust architecture of the unit with the better feature extraction ability of efficientNet-B7 can offer high accuracy and efficiency in wild boar recognition and segmentation. Figure 13.1 describes the unit architecture.

13.3.1 Unet

The UNet architecture is a convolutional neural network (CNN) specifically developed for semantic segmentation tasks, but better suited to the applications within the medical imaging analysis field. It consists of a contracting path, known as the encoder, while the other is the expanding path, or decoder. The primary function of the name encoder is to obtain the contextual information and features from a given input image. The function of the decoder is to provide the spatial detail and the segmentation map. For this reason, the network can calculate the exact positional prominence of the items within the image at a given time.

The specifics of the layers incorporated in the UNet architecture:

- **Encoder Layers**

  The first layer of the UNet architecture is the encoder, which consists of a series of convolutional layers and pooling layers. They are made to draw highly complex features from the input picture while

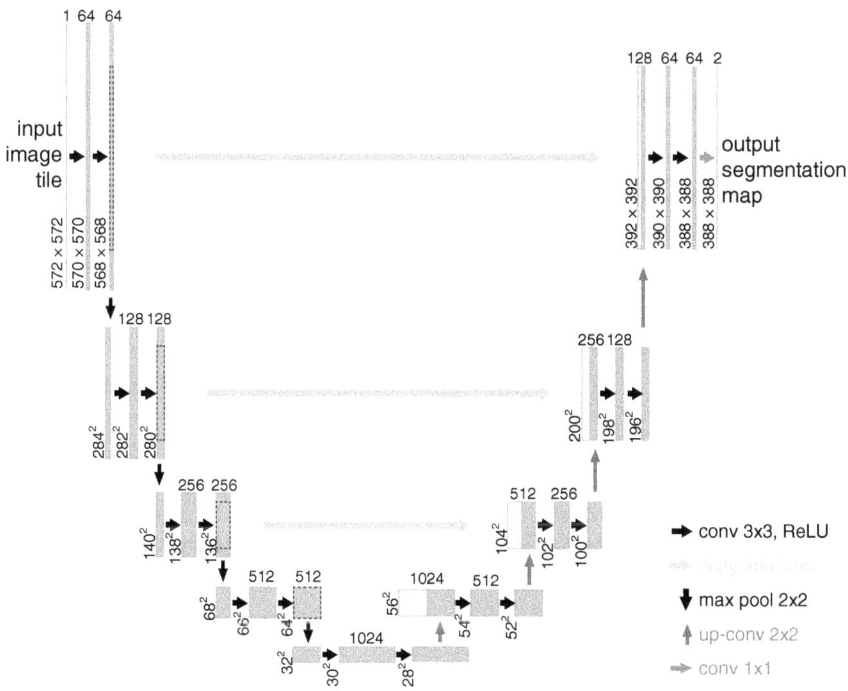

Figure 13.1 Unit architecture.

reducing its spatial size. The convolutional layers utilize filters to pull features from the digitized input image, whereas the pooling layers decrease the spatial size while maintaining important information about the input feature. The downsampling procedure aids in capturing contextual information for broader receptive fields.

- **Bridge Layer**

  After several rounds of downsampling during the encoder route, the most common practice is to add a bridge layer that serves as a connection between the encoder and the decoder pathways. The bridge layer is created to maintain a high-detailed representation of the features captured by the encoder as it moves to the decoder and helps localize accurately. Usually, it has additional convolutional layers that refine the captured features.

- **Decoder Layers**

  The decoder of the UNet is designed to increase the resolution of the encoder's obtained features to produce the final segmentation map. The decoder is designed as a set of upsampling layers that are then followed by layers of convolution. The dimensions of the feature maps in terms of space are increased gradually in these layers, where the spatial information is not lost. The upsampling layers would increase the original input image's resolution feature map match. This allows for better localization of the objects and borders within the segmented image.

- **Skip Connections**

  The critical innovation from the UNet architecture is the introduction of skip connections that join the corresponding levels of the encoder and decoder paths. Skip connections allow passing the exact spatial information from the encoder to the decoder, allowing the fully connected layer to recover fine detailed features lost in the downsampling phase. The UNet architecture accurately localizes objects in the segmented image by selecting features of various resolutions, using the encoder context as a reference for matching.

- **Output Layer**

  The output layer is the termination of the output route and is positioned at the end of the decoder. It is a convolutional layer programmed with a convolutional operation and activated with a sigmoid or a softmax function as its primary nature. The segmentation map is delivered as a pixel label for the entire input picture, enabling pixel labeling to recognize regions of interest in the input, like pictures, medical building instruments, and more.

  To conclude, a UNet comprises the Encoder, Bridge, Decoder, Skip connections, and output layers. Each one is critical in the field of feature extraction, downsampling, upsampling, and localization, which is valid for any segmentation work. As a result, it remains the best model in different works, especially in biomedical imaging. In addition, in corporate activities that demand the perfect anatomy and tissue demarcations to be generated to help process interventions, such an architecture is heavily used.

Advantages of UNet

- **Architecture in the Form of U:** The UNet design consists of a contracting route, which is composed of downsampling, followed by an expanding path, which is composed of upsampling. The architectural design portrays the shape of the letter "U". This enables the accurate identification of specific characteristics, thus creating detailed divisions or segments.
- **Skip Connections:** Within UNet, equivalent layers in the contracting and expanding routes are linked through skip connections. Skip connections enable the network to retain high-resolution information generated by the contracted route, facilitating the reconstruction of spatial information discarded during downsampling.
- **Reduced Vanishing Gradient Problem:** The UNet's skip connections help reduce the vanishing gradient problem, a common problem faced by deep neural network models during training. These skip connections establish direct links between the initial and ending layers, which improves the flow of gradients across the network layers, encouraging quicker convergence and more accessible training.

- **Efficient Use of Training Data:** The UNet optimally utilizes the training data through data augmentation techniques and transfer learning, which decreases the number of annotated training samples demanded compared to other segmentation architectures. It is beneficial when the labeled data is scarce or financially expensive.
- **Versatility in Picture Size:** UNet is flexible and can be trained with images of different sizes. Based on this, it is critical when it comes to working with medical data, as the images might contain significant disparities in terms of quality and size.
- **Cutting-Edge Performance:** UNet has achieved state-of-the-art results on a variety of medical image segmentation projects through training on a limited number of samples, including cell segmentation, organ segmentation, tumor identification, as well as lesion contours.
- **Rapid Inference:** The UNet architecture allows for speedy inference times, making it a good choice for any application that requires real-time or near real-time processing, for example, image guidance for surgery or when rapid decisions are needed, such as for diagnostics.
- **Versatility across Domains:** Originally, the UNet was developed to work on medical image segmentation, but it found applications in other areas, such as satellite image analysis, industrial inspection, and even autonomous driving. This serves as an indication of its effectiveness and potential for diverse applications.

In conclusion, the UNet architecture is a powerful technique for various segmentation operations in which the precision of object localization and segmentation in the picture is crucial. This is accomplished via a practical implementation of skip connections, one of the few designed to be U-shaped, and rapid training methods.

13.3.2 Efficientnet-b7

EfficientNet is a family of CNN designs designed to provide optimal performance with minimal computational cost. The label "EfficientNet-B7" is used to identify an individual member of this family by its exact depth, width, and resolution scaling parameter values. Figure 13.2 describes the MBConv block.

13.3.2.1 MBConv block

The MBConv block is a central building block of the EfficientNetB7 architecture, a state-of-the-art CNN designed for rapid and effective image classification use cases. Let's look at and scrutinize every part of the MBConv block:

Expand: This is the first step of the MBConv block operation. The number of channels or the existing feature mapping in the input tensor is increased. Typically, a 1×1 convolutional layer with many output channels is used to achieve this, compared to the output. The main aim of this expansion is to allow the model to capture a wide variety of diverse features from the input.

Depthwise Convolution: Following the expansion stage, the expanded tensor was subjected to a depthwise convolution, in which each port retained only one convolutional filter. During a standard convolutional layer, all filters must engage with all input ports on the way out. On the contrary, depthwise convolution employs just one filter for each port. This method substantially lowers the computation expense while effectively obtaining spatial characteristics.

Squeeze: A squeeze operation is applied after the depthwise convolution is executed. Usually, it operates through global average pooling, which covers the spatial information over all channels by computing the resulting average for each channel. The result is the tensor of reduced spatial dimensions and preserved per-channel knowledge.

Reshape and Reduce: After the squeeze stage, the output is then reshaped and reduced at this stage. The primary task during the reshaping process is to transform the tensor into a smaller size or reshape it into a new shape if possible. Following the reshaping process, the channels need to be reduced further. Typically, the reduction process involves 1 × 1 convolutions. This reduction enables the next stage to reduce the information while maintaining its high representational capacity.

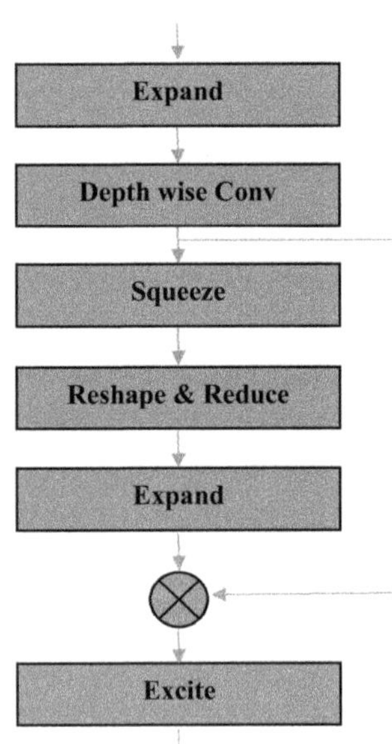

Figure 13.2 MBConv block.

Excite: The last operation in the MBConv block, the excitation procedure aims to boost significant characteristics while reducing less critical ones. This is done by frequently recalculating the feature maps to reweight them during inference dynamically. The process often necessitates the collection of attention weights for each channel and applying them separately to the feature maps to raise the importance of the dominant channels and minimize that of less essential channels. However, this operation's stimulation aids the model in focusing more on important characteristics and improves its separation capability.

In short, the MBConv block executes these operations in a well-designed order that effectively collects and cultivates unique features inside the input tensor. At the same time, it minimizes the computational expense and model parameters. Even if it has already become known as a highly successful design pattern in many picture classification undertakings, it hangs on to a high level of efficacy.

13.3.2.2 Architecture of efficientNetB7

The following Figure 13.3 describes the Architecture of EfficientNetB7

- **Input Layer:**

  The EfficientNet-B7 neural network's input layer is the first data layer, and the model's lead layer is concerned with how the raw input data, usually in the form of pictures, is fed into the ensuing layers of the model. What makes the input layer so critical is that it defines

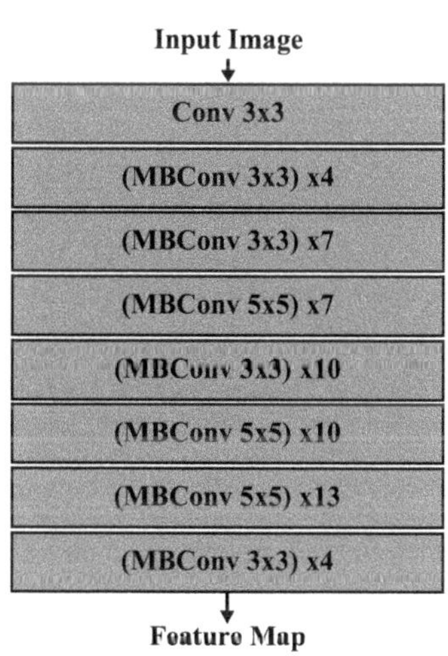

Figure 13.3 Architecture of EfficientNetB7.

the format and configuration of the input data the network expects to receive, whether in learning or performance. In contrast to other CNN designs, EfficientNet-B7 has a flexible input layer that takes pictures with disparate dimensions and aspect ratios.

In particular, EfficientNet-B7 uses AutoML, which is a method that automatically searches for the best-performing input picture dimension based on available machine computation resources for similar models. This process involves resizing the incoming photographs into various sizes and checking the MPAs produced by the resulting models. The final best-balanced accuracy/efficiency scale image resolution is chosen as the input picture dimension that the model expects to receive. Once decided, the input layer awaits pictures of that size, frequently expressed in height, width, and color channel number. For instance, if the dimensions for the input layer are set at 600 x 600p, the layer expects a 600p high and 600p broad picture, with pixels at three different color coordinates: red, green, and blue. During learning and performance, the input layer processes the input pictures by conducting preprocessing.

This process gets input pictures to fit most models' height, width, and color pixel integration settings. The method also includes scaling, normalization, and augmentation elements and is designed to help the model learn more robust and consistent features toward more accurate predictions. Therefore, the efficient net-b7 input layer is instrumental in enforcing the input data format, determining the ideal input picture dimension, and making the input pictures ready for easy learning by the rest of the network. Compared with traditional CNN designs, EfficientNet-B7 is highly adaptable and scalable, responsive to varied and asymmetrical input resolution, and runs more diverse picture sizes. Thus, it is ideal for most computer vision loads.

- **Stem Convolutional Layers:**

  The Stem Convolutional Layers of the EfficientNet-B7 architecture play a critical role in the initial pre-processing of input images. It serves as a gateway for the neural network, and its primary task is to extract the underlying patterns in input images while compressing them to simplify the analysis in the subsequent levels. The stem convolution layers comprise several convolution processes designed to identify the rudimentary elements, such as edges, patterns, and colors. It was necessary to lay the foundation for the network to progressively learn more complex and abstract patterns as the data traveled through multiple levels.

  The input image is subjected to convolution with filters at the initial stages. Present as small matrices, the filters run over the image and perform element-wise multiplication and summation. Since the process is done across all spatial positions, it helps identify patterns occurring at various places. Initially, convolutions have small design dimensions to detect complex elements present in input images. After the initial convolution processes, additional layers, such as pooling or striding, can be applied to reduce the feature maps' size in terms of spatial dimensions.

The purpose of downsampling is to reduce the computational load for the network, additionally boosting its receptive field, enabling it to capture more of the correlation and global trends in data. The design of stem layers on EfficientNet-B7, thus, is very well thought out, appropriately balancing computational efficiency and representation capacity. Most convolutional layers used in stem layers have reduced computational costs without much compromise on performance. Methods such as depthwise separable convolutions are used to factorize typical convolutions into independent operations for spatial and channel-wise dimensions.

- **EfficientNet Blocks (MBConv Blocks):**

  The EfficientNet Blocks, also known as MBConv Blocks, are crucial parts of the EfficientNet design, intending to generate a balanced model that combines efficiency and effectiveness. These blocks are essential for achieving state-of-the-art levels with far fewer parameters and computation costs than traditional CNNs.

  The MBConv block is achieved by using depthwise separable convolution. Depthwise separable convolution originates from the idea of separating the traditional convolution into two—the depthwise and the pointwise convolution layers. This separation makes it far more computationally efficient by reducing the number of parameters in the parameter space and computing the spending pool while preserving training capacity. Depthwise convolution is a stage in which convolution is separately conducted utilizing single filters supplied to each input channel of the feature map. It results in spatial data. Find the information from each solitary filter and each of the input channels. The computation and operation manage to be far cheaper than traditional input controls. In the subsequent phase, termed pointwise convolution, the channel count of the feature maps is increased, followed by the utilization of 1 × 1 convolutional filters on the entire input volume.

  The final step is the ReLU activation function, a linear activation used in contemporary neural networks. Increasing channels thus allows the network to understand more complex and interpretative features while maintaining computational efficiency. Additionally, MBConv Blocks frequently have extra modifications to aid feature representation further and improve the model file. A common addition to MBConv Blocks is using a squeeze-and-excitation block in each MBConv Block. This block assists by dynamically recalibrating the feature responses by learning to focus on informative feature channels. This is achieved by allowing the module to emphasize different feature planes, hence knowing how to talk to different kernels and layers with given weights.

- **Global Average Pooling Layer:**

  The Global Average Pooling Layer is a component found in many CNN architectures, including EfficientNet-B7. The primary purpose of this step is to convert the spatial data in the feature maps produced by the last convolutional layers into a compact form suitable for classification or regression. The feature maps produced by the latest convolutional layer contain much information regarding the distinct spatial patterns of the input image after many convolutions and activations. It is necessary to reduce the spatial dimensions of the feature maps before they can be piped into fully connected layers for classification, while maintaining the essential information. This drop can reduce the number of parameters in the succeeding layers, preventing overtraining and cutting the computational cost.

  The mean value of each feature map over its spatial dimensions is calculated to reduce the dimensionality of the feature maps in the Global Average Pooling Layer. Global average pooling is implemented differently from ordinary pooling layers, even though it computes both the max and average pooling over a predefined kernel size and stride. For each channel of the feature map, the global average pooling layer calculates the mean activation value by adding the activation values from all the spatial locations within the channel. It is divided by the total number of spatial locations.

  The outcome is a feature vector in which each element is equivalent to the average activation value of specific feature map channels. Global average pooling calculates the mean value of its feature map's overall spatial dimensions. Such a pooling operation will include only critical information and rarely consider inconsequential geographical data. The pooling method suggested in this concept is highly beneficial because it maintains the geographical context while significantly diminishing the feature representation's dimensionality. These properties allow for better computational efficiency and less overfitting.

  Furthermore, enhancing the network's ability to remain invariant to translations through the operation of a global average pooling layer reduces the dependence of the output of this layer on the exact position of the features in the feature maps. This addresses the network's ability to translate, rotate, and distort the input image in its ability to learn and improve its performance.

- **Fully Connected (Dense) Layer:**

  Fully Connected Layer, known as the Dense layer in neural network topologies. Its central role is to capture intricate patterns from the information aggregated by its precursors. This is the high-level reasoning component of the network; it is critical for classification, regression tasks, and image recognition. A fully Connected Layer is one where every neuron in the current layer is connected to every neuron in the next layer, producing the dense matrix. It is so densely connected that it allows capturing nuanced, practical dependencies between features. Therefore, the network can better understand complex patterns.

In the EfficientNet-B7 design, the Fully Connected Layer is often fed by the Global Average Pooling Layer. After global average pooling, the feature maps are flattened to a one-dimensional array or list, used as input for the dense layer. Every neuron in the densely connected layer makes an individual computation within the layer in mathematics. Every neuron in the thick layer multiplies its inputs by its corresponding weight, sums them all together, and then all are passed through an activation function to bring the non-linearity into play. The weight and bias used for each neuron are tuned during the training by backpropagation. Backpropagation is an optimization technique for a given parameter. It prepares the network, which helps it modify them so that the discrepancies between the expected and actual output are minimized. Our network uses this phase to change its parameters and understand the underlying patterns in the training dataset.

In the case of a classification task, the Fully Connected Layer output is frequently passed through a softmax activation function. A softmax transforms the raw output values into probability distributions over the given number of classes. The network can generate predictions by selecting the class with the highest probability.

- **Output Layer:**

The last layer in a CNN is the Output Layer. The model offers predictions or outputs through these channels. The design and structure of the model should be used to achieve a specific purpose, such as classification or regression.

For classification, and since EfficientNet-B7 is good at image classification, the output layer generally has a dense layer followed by a softmax activation function. They are mainly used to convert the features extracted by the following layers to the class probabilities of the different classes. Every neuron in the output layer is associated with a given class. The softmax function is later used to convert the raw output scores for all classes into a standardized probability distribution, where the probability of each class is represented. The model can then relate the likelihood of input forms part of the class.

For example, if the model learns to classify pictures into 1,000 distinct classes in the image data, such as an image, the output layer contains 1,000 neurons, each of which forms a class. Softmax activation ensures that the sum of class probabilities equals 1, which makes it possible to understand how confident the model is in making the prediction. In a regression task, where the goal is to predict a continuous value, the output layer can contain a single neuron with a linear activation function. For example, if the task is to predict the house price based on the house features, the output layer provides a single output, which stands for the price.

It is also crucial to note that the output layer acts only to generate predictions. Still, the final results are obtained by compiling the information extracted from the rest of the previous layers, notably the convolutional layers, pooling layers, and fully connected layers. The output layer merely gathers these features and passes them through another transformation to make final predictions.

13.3.3 Proposed combination of Unet Efficientnet-b7

Integrating Unet and EfficientNet-B7 proposed for wild boar semantic segmentation is a new method in computer vision and image processing. Unit is a well-established architecture for semantic segmentation. It is popular because it creates accurate segmentation of images by semantically determining pixels. Structurally, it is a composition of an encoder-decoder, in which the original object sample extracts the local features, and the other one receives the global features. Such a structure is beneficial in the recognition and segmentation of objects. EfficientNet-B7 is a state-of-the-art CNN architecture. This model achieved substantial results in various image classification fields. Moreover, it is characterized as a model that is able to utilize the computational power most effectively while having a small model size. This is achieved by the uniform scaling method, which equally scales the network's depth, width, and resolution.

Combining the above two modules, Unet and EfficientNet-B7, can bridge the strengths of both architectures. The model proposed in this paper is expected to provide a more accurate and faster approach to semantic segmentation for wild boar. The EfficientNet-B7 was utilized to obtain the high-level features of the input images. In contrast, Unet was applied through the encoder-decoder path to enhance the features across the upper layers, resulting in an extraordinary semantic segmentation mask. Figure 13.4 shows the Architecture of the proposed UNet with EfficientNetB7.

- **Conv 2D**
  Convolutional layers are central to the CNN design, as they are specifically developed for feature extraction. To perform this function, they create filters processed with incoming data to reveal features through convolution. Filters shifted horizontally and vertically over small items of the inputs produce feature maps. I assume that "Conv 2D" suggests regular 2D convolutional layers that extract only the necessary feature data from the input image.
- **MBConv 3 × 3**
  MBConv, short for Mobile Inverted Bottleneck Convolutions, first emerged in the EfficientNet family. Core components of MBConv may be divided into three distinct stages: depthwise convolution, expansion convolution, and pointwise convolution. This MBConv 3 × 3 means the depthwise convolution layer used a 3 × 3 kernel for the feature convolutions. They have constructed the convolutional

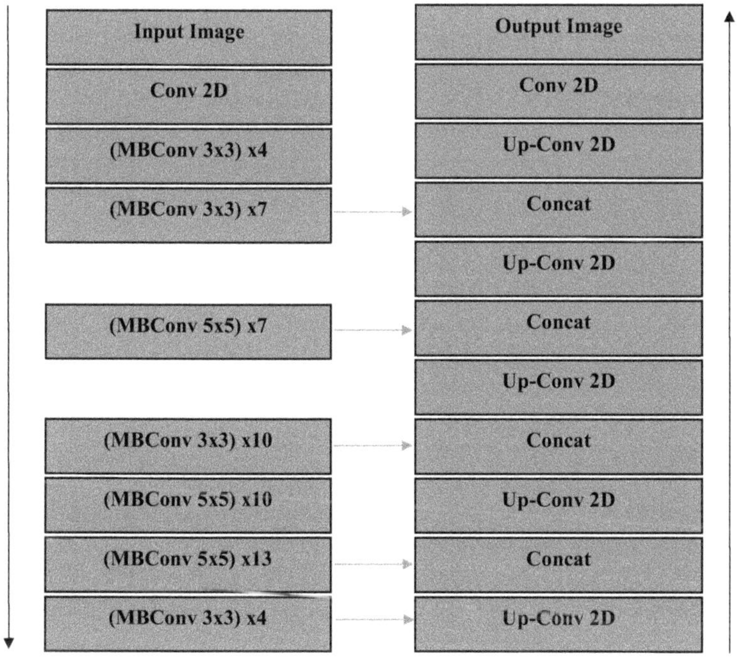

Figure 13.4 Architecture of proposed UNet with EfficientNetB7.

layers as a kernel that is either highly efficient in computing or very successful at learning complex patterns from data.

- **MBConv 5 × 5**

  MBConv with a 5 × 5 kernel utilizes a larger kernel size (5 × 5) for the depthwise convolution process, similar to MBConv with a 3 × 3 kernel. This modification enables the model to include broader spatial features in contrast to its 3 × 3 counterpart, perhaps enhancing its ability to detect patterns over more significant regions in the input data.

- **Up Conv 2D**

  Up-convolutional layers, also referred to as transposed convolutional or deconvolutional layers, are architectural components that have been developed explicitly for semantic segmentation. Up-convolution is the fundamental method of increasing the resolution of feature maps, thus allowing the network to predict the label for every pixel densely. According to this architecture, the operation is called "Up Conv 2D" and is meant to upsample. This is done to maintain the model's ability to determine the exact borders of a segmentation appropriately.

- **Concatenation**

  Concatenation is the process that is considered to be crucial for neural network topologies where models are created based on skip connections, such as for UNet and its variations. The process is combining feature maps by a few levels or branches of the network, and it is likely to assume that in the given architecture, the given term "Concat" is the integration process between feature maps received both on the contracting and the expanding route, based on the type of integration – concatenation. Multi-scale feature integration allows for dramatically better segmentation performance, as all the available information for all the possible input data scales is better utilized.

In conclusion, better results are attained when the expanding and contracting route features are combined at distinct scales. It can be said that the EfficientNetB7-based architecture employs several components—traditional convolutional layers, efficient MBConv blocks with different kernel sizes, up-convolutional layers for upsampling, and concatenation operations—to achieve good performance in semantic segmentation tasks.

13.3.4 CSRT tracking

Object tracking remains an essential problem in computer vision, and it has multiple uses, such as surveillance and augmented reality. The Continuously Adaptive Mean Shift tracking is a popular object-tracking method. It lays the foundation for the Channel, Spatial Reliability Tracker (CSRT) approach. The CSRT tracking algorithm is very robust in maintaining the view of its object in complex scenarios, such as occlusion, size variation, and lighting variations.

The CSRT tracking methodology revolves around the basic notion of describing the target object both in the spatial and feature domains. Their representation enables the tracking module to deal appropriately and efficiently with the behavioral variations and disappearance of the object. The method begins with choosing the target object in the first frame of the video sequence. The CSRT method represents the object of the human body through spatial and feature terms.

Specifically, the CSRT technique generates a spatial box around the human body, dependent on the initial position. Although the box is typically referred to as an area of interest for the following frames, the CSRT tracker does not keep a "box-based" representation. Alternatively, the CSRT tracker maintains the spatio-temporal data of the human body. To do this, the CSRT method introduces a dense sampling scheme inside the area of interest, which is the subsequent representation of the box in the tracking framework. The goal is that this sampling approach enables the CSRT tracker to adequately adapt to object-spatial adjustments, such as placement and size.

Furthermore, the CSRT approach not only maintains the spatio-temporal data of the object of interest but also permits the target object's shrink presentation. The partial representation enables the CSRT to tackle changes in appearance because the CSRT is concerned with robustness.

The CSRT technique uses advanced algorithms like the histograms of oriented gradients and the color histograms to represent the object of interest faithfully. This permits the CSRT to conserve robustness during variations in the appearance of the object target due to changes in lighting, among other issues.

After the first model of the target item is created, the advanced search algorithm of the CSRT helps to locate the object very accurately in upcoming frames. The search is conducted based on spatial and feature-based inputs, which help to adapt vehemently to the changes in the object's position, size, and specific appearance. The spatial reliability factor ensures that the tracker mainly concentrates on the vicinity of the prior object position. In contrast, feature-based reliability directs attention to the areas with qualitatively identical visual characteristics to the target item.

During the tracking, the CSRT algorithm continuously improves its representation of the object through the observed data from every frame. This updating approach helps the tracker adjust to the changes in the object's appearance and movement pattern and ensures highly performant and reliable functioning throughout extended tracking. Additionally, the CSRT has mechanisms that help address occlusion issues by dynamically changing the tracking window or reinitializing the tracker.

Ultimately, the CSRT tracking method uses spatial and feature-based object representations, resulting in robust and reliable performance in visually complex settings. CSRT is extremely precise across various tracking workflows due to high-detailed spatial sampling, discriminative feature extraction, and advanced search. Hence, this approach can significantly enhance many computer vision efforts.

Steps for a CSRT tracking algorithm:

Algorithm: CSRT tracking

Step 1: Begin by initializing the tracker with the first frame.

Step 2: Derive characteristics from the first frame

Step 3: Enhance the model by including the retrieved characteristics

Step 4: The position of the item is projected for each successive frame.

Step 5: Features are extracted from the anticipated area.

Step 6: Compute the confidence score for the projected location

Step 7: If the confidence score exceeds the threshold, update the tracker with the new location.

Step 8: Repeat steps 4–7 for every frame in the sequence.

13.4 EXPERIMENTAL RESULTS

This section provides a comprehensive analysis of the simulations' results using the recommended methodology. This study was carried out with the help of the datasets that contain images and videos of wild boar.

The training loss measures how far the model's predicted output is compared to the actual target data while training the model, as shown in Figure 13.5. It calculates how well the model performs on the training data. The training loss reduction is achieved using an optimization method like gradient descent. These optimization algorithms change model parameters multiple times to minimize the loss of the training data. Unlike the training loss, the validation loss calculates how well the model works on unseen data, or the one not used for training.

The unseen data, also known as the validation set, represent real-world data the model has never seen. They are used to determine if the model can generalize predictions to new data not seen before. Therefore, the validation loss is a measure of how the model can generalize behaviors learned from training to new, unseen data.

The training loss and the validation loss are monitored during training. The purpose is to minimize the training loss to achieve a high score based on the training data and avoid overfitting the model. Overfitting is encountered if the model memorizes the training data and does not base it on the actual pattern, leading to a lousy model behavior while predicting new

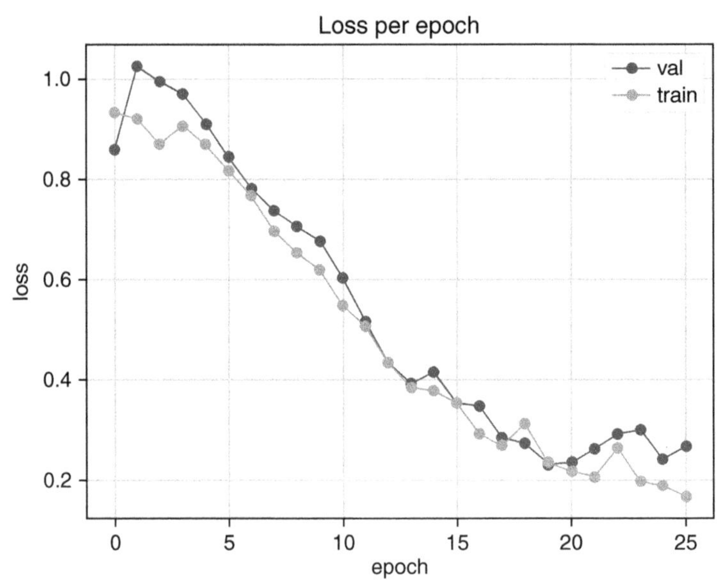

Figure 13.5 Train loss vs. validation loss.

data. Validation loss is used to evaluate overfitting: if the validation loss is increased and the training loss is decreased, the model is overfitting.

In the domain of computer vision, Mean Intersection Over Union (mIoU) is an essential parameter to determine the semantic segmentation's quality. This metric uses the statistical method to compare the predicted segmentation mask with the ground truth mask by calculating the intersection over union (IoU) for all the classes and averaging them, as shown in Figure 13.6. IoU is a measure of visible overlap that quantifies the proportion of commonality between a predicted mask and ground truth masks with respect to IoU. The metric shows how effective the model can be when it's about predicting an item's border. Hence, utilizing the formula for mIoU, we can derive the IoU scores of every class and consolidate them. The calculation will yield a metric where the parameter examines all outputs deeply, and the better output will have a higher value.

Training accuracy, which is how well a model performs on the data it is trained on, is shown in Figure 13.7. The model adjusts its parameters throughout the training process until the predictions made by the model closely match the actual target/value in the training data. As the training progresses, the model becomes more skilled at accurately depicting the patterns present in the data, which leads to increased precision. A high training accuracy, in turn, means that the model has managed to learn the patterns in the training data effectively; however, in some cases, it does not guarantee performance on unknown data.

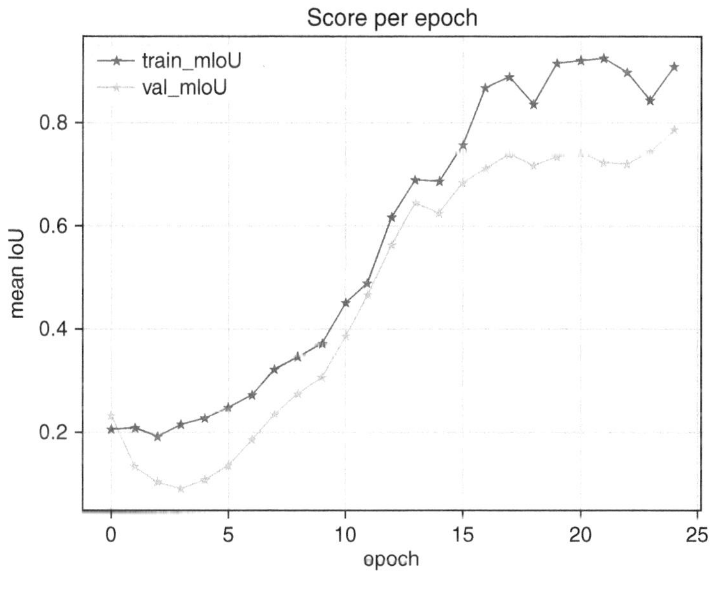

Figure 13.6 Train mIoUVs validation mIoU.

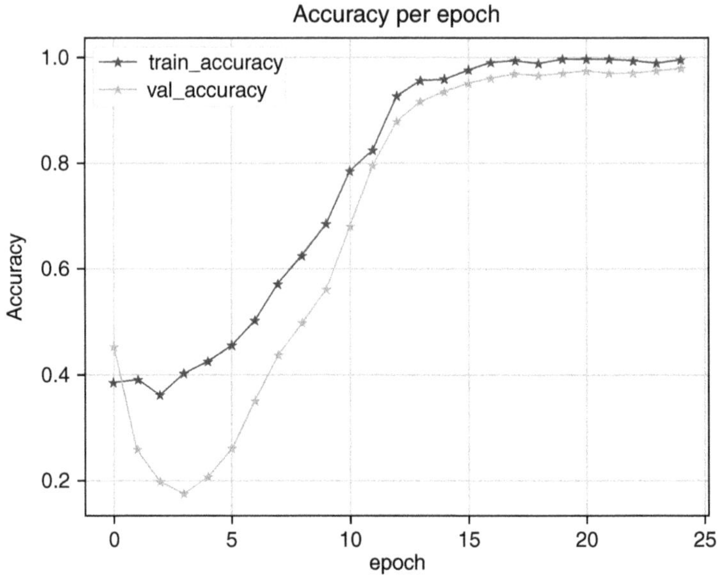

Figure 13.7 Train accuracy vs validation accuracy.

Validation accuracy is how well the model does on another dataset in training called the validation set. The validation set is made up of the samples in our dataset that were not included in training. The purpose of a validation set is to fairly assess how well a model has learned to generalize from unfamiliar situations. If the model makes accurate predictions with the unknown data, it has apparently learned the underlying concepts inherent in the data without focusing on the patterns specific to the training examples. However, if the accuracy of the validation is low, the model has overfitted. Overfitting means the model has memorized the training data rather than learning it and cannot generalize to new, unseen data.

13.4.1 Segmentation

The segmented image using the proposed method is illustrated in Figure 13.8. Figure 13.8a and b represent the input image and ground truth of the input image; on the other hand, Figure 13.8c shows the segmented image for the input image.

Table 13.1 displays a comparative examination of several techniques for picture segmentation using the EfficientNet-B7 architecture. The evaluation includes four methods: MAnet, FPN, UnetPlusPlus, and the proposed UNet architecture. Accuracy and mIoU serve as performance measurements. MAnet has a precision rate of 92.78% and a mIoU of 60.91%, while FPN exhibits a superior precision rate of 97.01% but a lower mIoU of

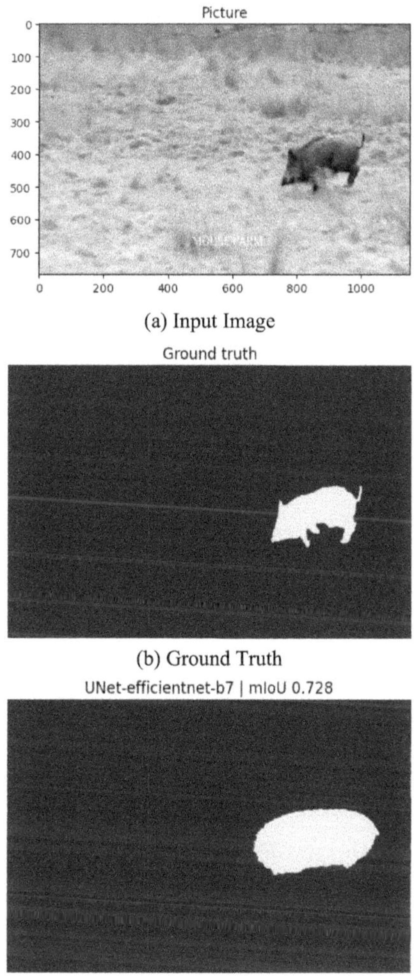

(a) Input Image

(b) Ground Truth

(c) Unet-Efficientnet-b7

Figure 13.8 Output segmented image.

Table 13.1 Comparative analysis

Method	*Accuracy*	*mIoU*
MAnet-efficientnet-b7	0.9278	0.6091
FPN-efficientnet-b7	0.97013	0.4850
UnetPlusPlus-efficientnet-b7	0.9281	0.6066
UNet-efficientnet-b7 (Proposed)	0.97032	0.7283

48.50%. UnetPlusPlus demonstrates comparable performance to MAnet, with an accuracy of 92.81% and a mIoU of 60.66%. Nevertheless, the suggested alteration to the UNet design surpasses all other approaches with a precision of 97.03% and mIoU of 72.83%. This indicates that the suggested improvement greatly improves the segmentation performance in comparison to current approaches.

13.4.2 Tracking

This section describes the tracking results of wild animals using the proposed model. In order to validate the proposed model, three different video footages are considered. In these four different video footages, four different frames (images) are applied as input to the proposed model.

Figure 13.9 represents the tracking results of video footage-1 using the proposed model. Figure 13.9a–d indicates the input images of video footage-1, and Figure 13.9e–h indicates the corresponding tracking output images.

Similarly, the proposed model was tested on video footage-2, and the tracking results of video footage-2 are depicted in Figure 13.10. Figure 13.10a–d indicates the input images of video footage-1, and Figure 13.10e–h indicates the corresponding tracking output images.

The proposed model was also tested on video footage-3, and the tracking results of video footage-3 are shown in Figure 13.11. Figure 13.11a–d indicates the input images of video footage-1, and Figure 13.11e–h indicates the corresponding tracking output images. In each video footage, the proposed model exhibits better tracking results, as shown in Table 13.2.

CONCLUSION

The paper presented a groundbreaking method for the semantic segmentation of wild boars in agricultural settings by combining Unet and EfficientNet-B7 architectures. Overall, the proposed new integration has been carefully developed in the presented study to determine a new sophisticated approach for the challenging issue of detecting and monitoring wildlife encroachment on farmlands. Due to the ongoing careful experimental investigation, the proposed institution offered a phenomenal achievement in terms of accuracy and mIoU metrics, which reached 97.03% and 72.83%, respectively. Consequently, the proposed institution reveals the best performance in terms of segmentation efficiency compared to MAnet-efficientnet-b7, FPN-efficientnet-b7, and UnetPlusPlus-efficientnet-b7. In order words, the impact of the proposed integration should be regarded as a significant breakthrough on the use of farmland wild boar segmentation focus.

(a) Input Image-1 (b) Input Image-2

(c) Input Image-3 (d) Input Image-4

(e) Output Image-1 (f) Output Image-2

(g) Output Image-3 (h) Output Image-4

Figure 13.9 Tracking results of video footage-1.

Conflict of Interest: *The authors declare that there is no conflict of interest among them.*

Data Availability: *The dataset utilized in this study will be made available on request by contacting the second author of the paper via email.*

(a) Input Image-1

(b) Input Image-2

(c) Input Image-3

(d) Input Image-4

(e) Output Image-1

(f) Output Image-2

(g) Output Image-3

(h) Output Image-4

Figure 13.10 Tracking results of video footage-2.

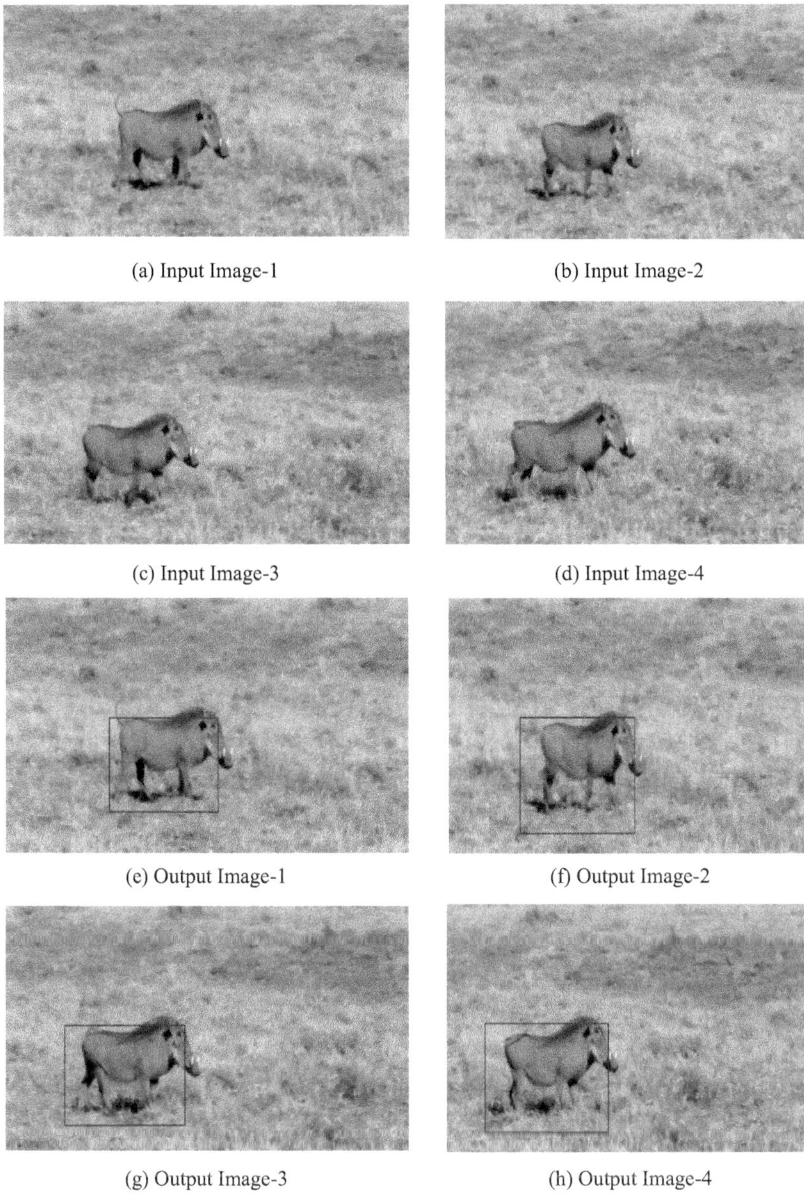

Figure 13.11 Tracking results of video footage-3.

Table 13.2 Tracking comparative analysis.

Algorithm	*Accuracy*
BOOSTING Tracker	88%
MIL tracker	89%
CSRT tracker	92%

REFERENCES

[1] Negi, Meenakshi, Mrinalini Goswami, and Sunil Nautiyal. "Agricultural loss due to wildlife: a case study on elephant depredation in a protected area of South India." *Journal of Social and Economic Development* 25, no. 2 (2023): 350–364.
[2] Yazezew, Dereje. "Human-wildlife conflict and community perceptions towards wildlife conservation in and around Wof-Washa Natural State Forest, Ethiopia." *BMC Zoology* 7, no. 1 (2022): 53.
[3] Carpio, Antonio J., Marco Apollonio, and Pelayo Acevedo. "Wild ungulate overabundance in Europe: contexts, causes, monitoring and management recommendations." *Mammal Review* 51, no. 1 (2021): 95–108.
[4] Zwerts, Joeri A., P. J. Stephenson, Fiona Maisels, Marcus Rowcliffe, Christos Astaras, Patrick A. Jansen, Jaap van Der Waarde et al. "Methods for wildlife monitoring in tropical forests: Comparing human observations, camera traps, and passive acoustic sensors." *Conservation Science and Practice* 3, no. 12 (2021): e568.
[5] Mekonen, Sefi. "Coexistence between human and wildlife: the nature, causes and mitigations of human wildlife conflict around Bale Mountains National Park, Southeast Ethiopia." *BMC Ecology* 20, no. 1 (2020): 51.
[6] Prasad, Tharun, and C. Lakshmi. "Coco-based crop protection system with integrated buzzer." In *2023 9th International Conference on Smart Structures and Systems (ICSSS)*, pp. 1–8. IEEE, 2023.
[7] Sayagavi, Ashwini V., T. S. B. Sudarshan, and Prashanth C. Ravoor. "Deep learning methods for animal recognition and tracking to detect intrusions." In *Information and Communication Technology for Intelligent Systems: Proceedings of ICTIS 2020, Volume* 2, pp. 617–626. Springer Singapore, 2021.
[8] Patel, Rishi Sanjaykumar, Dhrumi Devendra Amin, Sachin Patel, MrugendrasinhRahevar, Chirag Patel, and Amit Nayak. "Smart crop protection against animal encroachment using deep learning." In *2023 4th International Conference on Intelligent Engineering and Management (ICIEM)*, pp. 1–6. IEEE, 2023.
[9] Bhagabati, Bijuphukan, Kandarpa Kumar Sarma, and Kanak Chandra Bora. "An automated approach for human-animal conflict minimisation in Assam and protection of wildlife around the Kaziranga National Park using YOLO and SENet Attention Framework." *Ecological Informatics* 79 (2024): 102398.

[10] Rančić, Kristina, Boško Blagojević, Atila Bezdan, Bojana Ivošević, Bojan Tubić, Milica Vranešević, Branislav Pejak, Vladimir Crnojević, and Oskar Marko. "Animal detection and counting from UAV images using convolutional neural networks." *Drones* 7, no. 3 (2023): 179.
[11] Mao, Axiu, Endai Huang, Xiaoshuai Wang, and Kai Liu. "Deep learning-based animal activity recognition with wearable sensors: overview, challenges, and future directions." *Computers and Electronics in Agriculture* 211 (2023): 108043.
[12] Fernando, W. P. S., I. K. Madhubhashana, D. N. B. A. Gunasekara, Y. D. Gogerly, Anuradha Karunasena, and Ravi Supunya. "Image processing-based solution to repel crop-damaging wild animals." In *International Conference on Intelligent Sustainable Systems*, pp. 1–16. Singapore: Springer Nature Singapore, 2023.
[13] Natarajan, B., R. Elakkiya, R. Bhuvaneswari, Kashif Saleem, Dharminder Chaudhary, and Syed Husain Samsudeen. "Creating alert messages based on wild animal activity detection using hybrid deep neural networks." *IEEE Access* 11 (2023): 67308–67321.
[14] Sajithra Varun, S., and G. Nagarajan. "DeepAID: a design of smart animal intrusion detection and classification using deep hybrid neural networks." *Soft Computing* (2023): 1–12.
[15] Sayagavi, Ashwini V., T. S. B. Sudarshan, and Prashanth C. Ravoor. "Deep learning methods for animal recognition and tracking to detect intrusions." In *Information and Communication Technology for Intelligent Systems: Proceedings of ICTIS 2020, Volume* 2, pp. 617–626. Springer Singapore, 2021.
[16] Thomas, Aby K., P. Poovizhi, M. Saravanan, and K. Tharageswari. "Animal intrusion detection using deep learning for agricultural fields." In *2023 5th International Conference on Smart Systems and Inventive Technology (ICSSIT)*, pp. 1021–1027. IEEE, 2023.

Index

Note: **Bold** page numbers refer to tables; *italics* page numbers refer to figures.

For Product Safety Concerns and Information please contact our EU
representative GPSR@taylorandfrancis.com
Taylor & Francis Verlag GmbH, Kaufingerstraße 24, 80331 München, Germany

www.ingramcontent.com/pod-product-compliance
Ingram Content Group UK Ltd.
Pitfield, Milton Keynes, MK11 3LW, UK
UKHW020156260926
475702UK00047B/475
* 9 7 8 1 0 3 2 7 9 9 6 7 4 *